MOUNT EVEREST

1938

Printed in Great Britain at the University Press, Cambridge
(Brooke Crutchley, University Printer)
and published by the Cambridge University Press
(Cambridge, and Bentley House, London)
Agents for U.S.A., Canada, and India: Macmillan

Looking westward from Camp V (25,600 ft.): (Pumori prominent in middle distance). Early onset of monsoon clouds. (Panoramic with Pl. 24) (p. 82)

MOUNT EVEREST
1938

BY

H. W. TILMAN

CAMBRIDGE
AT THE UNIVERSITY PRESS
1948

Still, I think the immense act has something about it human and excusable; and when I endeavour to analyse the reason of this feeling I find it to lie, not in the fact that the thing was big or bold or successful, but in the fact that the thing was perfectly useless to everybody, including the person who did it.

G. K. CHESTERTON

PREFACE

THIS account of the 1938 Mount Everest Expedition is published with the consent of and on behalf of the Mount Everest Committee, but for the views herein expressed the author alone is responsible. Some questionable publicity recently given to new methods to be applied to the problem of Mount Everest points to the need for the plea which is made here for sanity and a sense of proportion.

The plates are from photos taken by members of the party. I have not attributed them individually, but the majority, and certainly those of outstanding merit, may be safely regarded as those of Mr F. S. Smythe.

I have to thank Dr T. G. Longstaff (President of the Alpine Club) and Dr R. J. Perring of Ryton for reading and criticizing the first draft and R. T. Sneyd, Esq. for correcting the proofs. The maps are copies of those published in the first place by the Royal Geographical Society.

H. W. T.

WALLASEY
February 1947

CONTENTS

ILLUSTRATIONS

PLATES

MAPS

CHAPTER I

INTRODUCTORY

The sight of a horse makes the wayfarer lame. BENGALI PROVERB

THE last book written about Mount Everest by Mr Ruttledge, the leader of the 1933 and 1936 expeditions, was aptly named *The Unfinished Adventure*. This present account should be read merely as yet another chapter in this adventure story, possibly one of those duller chapters from which even the best of adventure stories are not always free. In the twenty-five years which have elapsed since the first expedition went out the story has lost the gloss of novelty. The approach march and the establishing of camps have become almost a matter of routine which with luck and judgement should be devoid of incident. Misfortunes and hair-breadth escapes, suffering and hardship, are the making of an adventure story, but from all such a well-found expedition blessed with a fair share of luck should be exempt. Here I have no hardships to bemoan, no disasters to recount, and no tragedies to regret.

Some day, no doubt, someone will have the enviable task of adding the last chapter, in which the mountain is climbed, and writing 'Finis'. That book, we may hope, will be the last about Mount Everest, for we already have five official accounts, besides a few unofficial, and no one can tell how many more will be written before the epic is complete. Apart from reasons of continuity in the record of this unfinished adventure, the story of the fifth abortive attempt to climb the mountain is only worth relating because a fairly drastic change was made in the methods used. That is to say we broke away from the traditional grand scale upon which all previous expeditions had been organized, and to that extent the story has novelty. But we made no change in the route taken or the tactics employed on the mountain, which are the outcome of the judgement and hard-won experience of some of the best mountaineers of recent times, whose achievements are a guide and an inspiration to all who follow where they led.

It is difficult to measure that margin in terms of additional effort (it may be greater than we think), but in view of the apparently narrow margin by which two of the earlier expeditions failed, it may seem presumptuous to imagine that any change of organization should be needed. So before recounting our experiences of 1938 I feel it is due to those who sponsored the expedition, the friends who backed it, and to the many mountaineers interested who may sympathize with some of the views here expressed, to attempt some explanation. The expeditions of 1924 and 1933 seemed to come so near to success that few if any thought of questioning the soundness of the methods employed, at least for the getting of someone to the top of the mountain; for long before then mountaineers had begun to dislike the excessive publicity which was a direct consequence of the scale of the expeditions and the large amount of money needed to pay for them. But after 1933 criticism began to be heard—Mr E. E. Shipton was possibly one of the first to doubt that in mountaineering the great and the good are necessarily the same—and the unfortunate experiences of 1936 when, through no fault of those concerned, but little was accomplished, had the salutary effect of rousing doubts in others. What had happened once might happen again. For financial reasons, if for no others, it seemed the time had come to give less expensive methods a trial.

Although our expedition of 1938 was the seventh to visit the mountain it was only the fifth to attempt the ascent. The first, and in many ways the most interesting, expedition was the reconnaissance of 1921 during which, of course, no attempt was made on the summit. Until 1921 no European had been within 90 miles of the mountain and the first party had to find the best approach and then a likely route to the top. Both these difficult tasks and much additional work were successfully accomplished at a cost of about £5,000—a figure which is not unreasonable considering the complete lack of previous experience, the time spent in the field, and the amount and importance of the work done. But the first attempt on the summit which took place the following year cost more than twice as much, and set standards in numbers, equipment, and cost, which until 1938 were equalled or even exceeded by all subsequent expeditions excluding only that most interesting and significant expedition of 1935 which was again a reconnaissance.

Late in 1934 the Tibetan Government unexpectedly announced that they would allow us to send an expedition in each of the following years, 1935 and 1936. Time was short, for in those days the gestation period for a full-blown expedition was, suitably enough, like that of a whale or an elephant, about two years; but so that the benefit of the surprising gift of the extra year should not be lost, Mr Shipton was hastily appointed to organize and lead a small, light expedition in 1935. Their main task was to try out new men and equipment for the full-scale attempt the following year; other tasks were the examining of snow conditions on the mountain during the monsoon and the survey of glaciers north and east of the mountain. At a cost of only £1,500 a large area of country and the North Face of the mountain were surveyed, and twenty-six peaks of over 20,000 ft. were climbed. In the course of these operations the North Col (Camp IV) was occupied, and it became plain that, had conditions warranted and had a few more tents been available, then a serious attack on the summit could well have been launched. This should have opened every one's eyes, especially as the expedition had been sent out so that its lessons might be of use to the all-out attempt of the following year. But this example of what could be done with a moderate expenditure was ignored and the expedition of 1936 saw no diminution in scale, either of men or of money. Twelve Europeans, including two doctors, a wireless expert, over a hundred porters, three hundred transport animals, and some £10,000 were employed, and the North Col was the highest point reached.

It is not easy to see either the origin of or the reasons for these unwieldy caravans organized on the lines of a small military expedition rather than a mountaineering party. Were it not that the pioneering days of Himalayan climbing were past one might find a parallel in the earliest days of mountaineering in the Alps, when numbers were considered a source of strength and not the weakness they usually are. For de Saussure's ascent of Mont Blanc in 1787 the party numbered twenty. The elaborately organized expedition of the Duke of the Abruzzi to the Karakoram in 1909[1] was the original Himalayan expedition in the grand style, but before and since that time many private parties had climbed and explored with a minimum of fuss and expense—notably those of Mummery,

[1] 270 persons (12 Europeans) and 95 baggage animals crossed the Zoji La.

Conway, Longstaff, Kellas, Meade, to mention a few.[1] Of course the means must be proportioned to the end; there is a difference between rushing a moderate-sized peak and besieging one of the Himalayan giants, but any additional means we think we need for the more formidable task ought to be taken reluctantly and after the severest scrutiny. Anything beyond what is needed for efficiency and safety is worse than useless. In 1905 Dr Longstaff and the two Brocherel brothers, with no tent and one piece of chocolate, very nearly climbed Gurla Mandhata, a peak in Tibet north of Garhwal, 25,355 ft. high, a practical illustration of the application of that important mountaineering principle, the economy of force—an imperfect example, perhaps, because one might argue that with a tent and two pieces of chocolate they might have succeeded. But away with such pedantic, ungracious quibbles. Did not Mummery, who more than any one embodied the spirit of mountaineering, write: '...the essence of the sport lies, not in ascending a peak, but in struggling with and overcoming difficulties'?

Though all mountaineers will agree with Mummery, it is no use concealing the fact that most of us do earnestly wish to reach the top of any peak we attempt and are disappointed if we fail: especially with Mount Everest parties where the desire to reach the top is supreme. No one would choose to go there merely for a mountaineering holiday. It is not easy therefore to criticize men for taking every means which they consider will increase the chances of success. It is a matter of degree, and on any expedition, even the most serious, the tendency to take two of everything, 'just to be on the safe side', needs to be firmly suppressed, for a point is soon reached when multiplication of these precautions, either in men or equipment, defeats its purpose.

Owing to the frequency of Alpine huts the longest climb in the Alps requires no more equipment than can be carried on the climber's back; while for numbers, although two are adequate and move fastest, three are no doubt safer. Any additional members usually lessen the combined efficiency of the party. In the Himalaya the peaks are twice as high and the climber has to provide his own hut. The climbing of a peak of, say 21,000 ft., will require a tent

[1] The Norwegians, C. W. Rubenson and Monrad Aas, who nearly climbed Kabru (24,002 ft.) in 1907 must not be omitted.

of some sort to be taken up to at least 17,000 ft. From a camp at this height a peak of 23,000 ft. has been climbed (Trisul by Dr Longstaff), but most people would prefer to have a second tent at some intermediate point from which to start the final climb. Obviously for higher peaks more intermediate camps are required and it becomes necessary to employ porters to carry and provision them. These porters will mean other porters to carry up *their* tents and provisions, and so it grows snowball fashion until in extreme cases like that of Mount Everest you have to find food and accommodation for at least fifteen men at 23,000 ft. in order to put two climbers in the highest camp at 27,000 ft.

It should be clear that the fewer men to be maintained at each camp and the less food and equipment they need, the easier and safer it is for all concerned. I am not advocating skimping and doing without for the sake of wishing to appear tough, ascetic, sadistic, or masochistic, but for the reason that no party should burden itself with a man or a load more than is necessary to do the job. If this principle be accepted and applied all along the line from the highest camp to the starting-point—London—the more likely will the expedition be economical and efficient, in short a small light expedition. The unattainable ideal to be kept in mind is two or three men carrying their food with them as in the Alps. How far this can be done has not been discovered—probably not very far—and there is the complication of supporting parties which though desirable are perhaps not essential. If the highest party is unable to get down on account of bad weather, the party below is not likely to be able to get up to help them—as happened on Masherbrum in 1938. Support or not, the importance of not being caught short of food reserves in the highest camp is obvious.

Between the two wars many small private parties, refusing to be frightened by the portentous standard set by the Everest expeditions and strenuously maintained by German and French expeditions,[1]

[1] International Expedition to Kangchenjunga, 1930: Europeans, 13; porters, 300. German Expedition to Nanga Parbat, 1934: Europeans, 14; porters, 600. French Expedition to the Karakoram, 1936: Europeans, 11; porters, 500. In these foreign expeditions of which the above are fair samples the climbing party usually numbered eight or nine, the balance being scientists, doctors, photographers, secretaries (in one case), and British liaison officers to run the porter corps.

accomplished much in the Himalaya, demonstrating that for peaks up to 24,000 ft. nothing more was needed, and thus keeping alive the earlier simple tradition of mountaineering with which the big expedition is incompatible. This was readily accepted, but the question of whether for the highest peaks the grand-scale expedition was either necessary, efficient, or expedient, was debatable. Every one recognized that an extra four or five thousand feet in height necessitated more camps and more porters, and although Nanda Devi (25,660 ft.) had been climbed practically without the help of porters it was admitted that similar methods would not work on Everest—the difference of 3,500 ft. in height between the two mountains is no adequate measure of the difference in degree of accessibility of the two summits. But if the provision of two more camps entails more equipment and more porters, it need not entail a small army with its transport officers, doctors, wireless officers, and an army's disregard for superfluity.

Then if the big expedition is unnecessary, is it efficient? In view of what was accomplished by the four expeditions up to 1933 it would be impertinent to say they were inefficient, but I believe the same men could have done as much, perhaps more, at a quarter of the cost, using methods more in keeping with mountaineering tradition. It is possible to argue that with less impedimenta to shift, fewer porters to convoy, and fewer passengers to carry, the strength of the climbing parties when it came to the last push would have been greater than it was. Theoretically the extra efficiency of the large party consists in having a reserve of climbers to take the place of those put out of action by sickness or frost-bite. Of the first four attempts of 1922, 1924, 1933 and 1936, the numbers taking part were thirteen, twelve, sixteen and twelve respectively. From these have to be deducted the supernumeraries such as non-climbing leader, base doctor, transport officers, wireless expert, leaving an effective strength of eight, eight, ten and eight. In 1938 there were seven of us. But we were all climbers and we carried only one-fifth of the gear and spent only a fifth of the money of previous expeditions. The actual number of climbers taking part does not define the 'big' or the 'small' party, and on this point there is not much in it between the advocates of either. As the best number to take part in the final climb is two, and as the odds against favourable conditions continuing long enough to allow of more than one attempt are high, provision for two attempts is all we need consider. Two parties of two and two spare men, six in all, should therefore be enough. A party of eight has a reserve strength of 100 % which should satisfy the most cautious. Most expeditions have had the benefit of a nucleus of three or four men whose ability to go high had already been proved. But if a party was not so fortunate in this respect, as might well be, then allowance would have to be made for a possible failure to acclimatize. The same would apply to the porters, for owing to the war and the lapse of time very few if any of the old 'Tigers' will be available.

With experienced porters such as we had the necessary convoy duties could be done comfortably by six or seven Europeans. Indeed, there is more to be feared from underwork than from overwork on these expeditions. Far too many off-days are forced upon a party and much time is spent lying about in sleeping-bags. Mr Shipton, who has taken part in two of the large expeditions, has remarked that there is sometimes a grave risk of contracting bed sores. I am sure many other climbers will bear me out that the common effect of too many off-days is a feeling of deadly lethargy. The risk of serious illness developing is not great and is over-emphasized. The party are fit men when they start and presumably able to take care of themselves. Coughs, colds, and sore throats seem to be inseparable from a journey across Tibet in the early spring, but their effects are not very grave. Indeed, it seems better to face the possibility of serious illness than the certainty of having useless mouths to feed and men falling over one another for lack of work.

A method is expedient if it tends to promote a proposed object. Whether the methods of these earlier expeditions were the best for the climbing of the mountain is a matter of opinion—I have tried to show that they were not, insomuch as they were wasteful and cumbersome—but I think there is no doubt whatever that they were not those best calculated either to preserve the well-being and goodwill of the peoples of those countries through which the route of the expedition lay or to maintain the best interests of mountaineering. The first point was raised in several letters to *The Times* in 1936 of which I quote from two, written by men[1] with Himalayan experience:

Apart from obvious cumbersomeness and expense, the huge expedition suffers from other serious objections. The first lies in the demoralizing effect which those visitations have on the people of the villages, by upsetting their scale of economic values. The arrival of an army of porters led by Sahibs apparently possessing boundless wealth and wasting valuable material along the route, makes a most corrupting impression. I was conscious of this when staying at the charming village of Lachen in North Sikkim through which several large expeditions have passed. The occurrence of theft in the later Everest expeditions, so out of keeping with the Tibetan character, is probably another case in point. It would be a tragedy if the visits of climbers to the Himalayas were to destroy one of its greatest charms—namely the honest character of the inhabitants and their splendid culture.

And in another letter:

In discussing the demerits of large expeditions a point not yet touched on is the unbalancing effect of the passage of a large transport column on the economic life of the country through which it passes. The part of Tibet traversed by the Mount Everest expeditions is by Tibetan standards and in comparison with the northern deserts fairly fertile. But actually only just enough barley is grown or can be grown to pay taxes and to carry the population through the winter and spring. The flocks of sheep and yaks are just as large as the grazing will permit. When an Everest expedition comes with 300 animals and a horde of hungry porters, reserves of food are broken into and sold: while grass which should have fed Tibetan ponies goes into the stomachs of the visiting yaks. It is true that good silver, British Indian rupees, are given in exchange. But, as a headman remarked to us in 1935 just after receiving Rs. 200, what good will that silver do? It cannot buy more corn where there is no surplus, nor will it fertilize the pastures. But it will quite certainly cause the neighbouring headman to be jealous, and enduring quarrels may be started. Where the balance between production and consumption is already precarious and where there are no reserves to draw upon the effect of a large expedition is materially disastrous.

There is certainly something in this; though the people along the line of march who receive good money for services rendered, who 'win' a number of useful articles of various kinds, and who also have their ailments attended to gratis, might take a less gloomy view. The Tibetans are shrewd people and will not exchange food for money unless they see their advantage in it, much less if it spells

[1] Marco Pallis: Michael Spender (killed in an accident May 1945).

starvation. Anyhow, the Indian Government is alive to this danger and now only one of what are euphemistically described as 'major expeditions' is allowed to operate in any particular area at a given time. And if the interests of the local inhabitants suffer from the large expedition so do the strangers who come after, who find the market for goods and services in a very inflated state.

Whether we climb mountains for exercise, love of scenery, love of adventure, or because we cannot help it, every genuine mountaineer must shudder involuntarily when he sees anything about mountains in newspapers. As Mr Jorrocks, who consistently refused to be weighed, used to say, when asked to mount the scales, 'his weight was altogether 'twixt him and his 'oss', so is mountaineering altogether a private affair between the man and his mountain; the lack of privacy is disagreeable and particularly so if, as usually happens, the newspaper gets hold of the wrong mountain wrongly spelt, adds or deducts several thousands of feet to or from its height, and describes what the wrong man with his name wrongly spelt did not do on it. Most human activities, especially the more foolish, are regarded nowadays as news, so there is perhaps the less reason to expect mountaineering to be an exception; nevertheless, it is indisputable that it is the big expedition that has occasioned this news value, the reason, of course, being that to finance them recourse must be had to the newspapers. No one bothers about climbers at home or in the Alps unless they fall, or in the Himalayas unless some newspaper is paying for the story. Thanks to this publicity the interest taken in Everest expeditions is now world-wide; news about them is published whether authentic or not, whether paid for or not; it is therefore very difficult, quixotic in fact, to refuse an offer for the story when a refusal will make not a jot of difference to the sum total of publicity.

The effect of this on the climbers taking part, and on mountaineers generally, should be taken into account. It is considered an honour to be asked to join a Mount Everest expedition; but, human nature being what it is, the exaggerated glamour which now surrounds an expedition of this sort has made the competition for a place even keener, so that much canvassing takes place before the final selection and much heart-burning after. The chosen party finds itself burdened with unnecessary responsibilities; responsibilities to a committee, to a newspaper, or even to the nation as the 'pick of young British manhood', as one unfortunate party was described. A feeling that the eyes of England are upon you may be very bracing before a battle but is not conducive to sound mountaineering.

Finally, it was publicity which engendered a competitive spirit individually and nationally. I think it is true that the big German expeditions received financial as well as moral backing from their governments and certainly the Tibetans themselves are convinced that we are sent to climb Everest at the bidding of our Government to enhance national prestige. One result of this is that mountains tend to become national preserves and the Indian Government has thrust on them the thankless task of deciding whether a party from one nation should be allowed to attempt a mountain that has already been visited by that of another. A question which might easily be settled by a little consideration and co-operation amongst the climbers themselves. The evils are there for all to see, but how or whether they can be abated is less obvious. Probably the phase is only a passing one, born of an age of advertisement. Many mountaineers fervently hope that the big mountains like Everest, Kangchenjunga, K 2, and Nanga Parbat, will be climbed soon so that Himalayan climbing may regain the more normal atmosphere of Alpine climbing and cease to be a mere striving for height records. Whether these mountains are climbed or not, smaller expeditions are a step in the right direction which should make even parties attempting the very highest peaks less subservient to publicity than heretofore, if not quite independent of it. Much will have been gained by that. For men living in India there should be no financial difficulty; but in the nature of things most young climbers in England are not in a position to pay their full share of the expenses of a Himalayan expedition; although, be it noted, two men did find it possible to spend five months in the Himalaya at a cost of £143 each, including passage money both ways. With the money that has been spent in trying to climb Everest a fund could have been endowed, the income from which would have more than sufficed to send a party annually to attempt some Himalayan climb. However, such a fund might have done more harm than good; a man who is bent on getting to the Himalaya will find ways and means.

Books, though they endure a little longer, are a less baneful form of publicity than newspaper articles because few read them. 'No man but a blockhead', says Dr Johnson, 'ever wrote except for money', a remark which is quite true of the writers of Mount Everest books who wrote in the first place to defray the expenses and who must now write to preserve the continuity of the story. Unlike the desert and the sea, mountains have not yet found a writer worthy of them. Perhaps those who could have written in a way that would live have felt about books and publicity as Mallory felt. In his contribution to the 1922 expedition book, which was a hint of his ability in this respect, he makes an eloquent but unavailing plea for silence: 'Hereafter, of contemporary exploits the less we know the better; our heritage of discovery among mountains is rich enough; too little remains to be discovered. The story of a new ascent should now be regarded as a corrupting communication calculated to promote the glory of Man, or perhaps only of individual men, at the expense of the mountains themselves.'

But we protest too much; for at bottom does it matter what is written about that 'considerable protuberance' Mount Everest or any lesser mountain? Let man conquer (revolting word) this, that, or the other, and write volumes about having done so, nothing he does or says will tame the sea or diminish the glory of the hills. The sea, immense and romantic though it is, is a commercial highway; the desert, with toil and money, can be made to blossom; but the mountains, thank heaven, are a sanctuary apart. None of our tricks can change them, nor can they change that man who looks at them for their beauty, loves them for the way they infect and quicken his spirits, and climbs them for his fun. 'If there are no famous hills, then nothing need be said, but since there are they must be visited.'

But though the mountains cannot change, our approach to them may; and if in this chapter, and the last, the note of criticism and protest appears faintly querulous it must be attributed to a perhaps presumptuous jealousy for all that mountaineering and mountaineering tradition stands for. Mountains mean so much to so many, as sources of comfort and serenity, as builders of health and character, and as strong bonds of friendship between men of all kinds, that any tendency to diminish their might, majesty and power, should be resented and resisted. Well was it said, 'Resist the beginnings', and therefore these protests have been written in the hope that promoters of Himalayan expeditions will think twice about the use of innovations designed to soften the rigours of the game or lessen the supremacy of the mountains. Of the many strange tricks that man plays before high heaven that would be one of the strangest, one which if it did not make angels weep would strike moralists dumb, if our efforts to subdue the mightiest range and the highest mountain of all should be the means of losing us our mountain heritage.

There will be small danger of this happening if we do not treat our highest mountains too seriously, the attaining of their summits as the only end, and an attempt upon them as 'man's expression of his higher self'—whatever that may mean. When our forerunners were busy discovering the Alps, as we are now discovering the Himalaya, I feel sure they did not look upon themselves as so many bearded and be-whiskered embodiments of man's unconquerable spirit striving to attain the highest. In German accounts of Himalayan climbing between the wars one came upon this high-falutin' attitude, and occasionally in reading of Mount Everest one detected a portentous note, as in a dispatch from the front. As soon as we begin to talk or write thus about men and mountains we should remind ourselves of a remark of Chesterton, that cheerful apostle of common sense and paradox: 'Physical nature must be enjoyed not worshipped. Stars and mountains must not be taken seriously.' Which I take to mean that we may enjoy our mountains and love what we enjoy, keeping our passion for them this side idolatry, but that mountaineers and astronomers committing their follies shall be viewed by others with the indulgence customary towards foreigners, dons, and the eccentricities of genius, and by themselves with only the very mildest esteem.

CHAPTER II

PREPARATIONS AT HOME

What creates great difficulty in the profession of the commander of armies is the necessity of feeding so many men and animals. If he allows himself to be guided by the commissaries he will never stir, and all his expeditions will fail.
NAPOLEON

THE unfortunate experience which befell the 1936 expedition converted many to the idea of smaller, less expensive, expeditions—above all less expensive. Not that a small expedition would have fared any better in that year, but the defeat would have been less resounding. Here was an expedition on which no expense had been spared, organized and equipped in a way which was thought would ensure success, turned back by unfavourable conditions at only 23,000 ft. What had happened once might happen again; and now that it had been shown that there might very well be no story worth selling, the large sum required for a similarly elaborate expedition would in future be less easily found. When, therefore, it was known that the Tibetan authorities would permit another attempt in 1938, by general consent the expedition was organized on a more modest scale. The Tibetan authorities have been very indulgent towards us in the past and we must hope for a continuance of their goodwill in the future; but the obtaining of permission to go to Mount Everest is still one of the major difficulties in the way of climbing the mountain. I venture to think that had it lain in a more accessible part of the Himalaya and had more frequent attempts been possible it would have been climbed before now. I have been told by one who should know that so sacrosanct do they regard Mount Everest that asking the authorities at Lhasa for permission to climb it is like asking the Dean and Chapter for permission to climb Westminster Abbey. That may be so, but it is certainly not the impression one gets from the people who live almost under the shadow of the mountain—the people who, one might expect, would be the most sensitive to any hint of sacrilege. I mean the people of Rongbuk, particularly the abbot of the Rongbuk monastery who has seen all the expeditions pass and who welcomed us in a most friendly way, doing all in his power to assist us. Of course, it may be said, he is so serenely confident of the mountain's power to defend and, if need be, avenge herself, that he just treats us with the innate courtesy and friendliness common to Tibetans. Gathering information through an interpreter is unsatisfactory, but so far as we could understand, the monastery is not a very ancient foundation; founded and built, in fact, largely by the present abbot who is now a very old man. Nor could we discover whether it was built on its present site for the sake of proximity to the mountain or for some other mythological association.

Permission having been given, the Mount Everest Committee appointed a leader and gave him, as is usual, a free hand. It was with some diffidence that I accepted this responsible post, for amongst those whom I proposed to invite were men with a greater mountaineering experience than mine and particularly more experience of Everest. But I took comfort from the thought that with men like E. E. Shipton, F. S. Smythe, and N. E. Odell amongst the party, it would be my part to sit listening with becoming gravity to their words of wisdom, waking up occasionally to give an approving nod. In fact I should have a sinecure, as should be the case with the leader of any well-balanced climbing party. And so it was so far as making decisions on the mountain went, most of which were imposed on us, willy-nilly, by the weather.

A party of seven was finally made up, of whom only one was not a first choice. It was unfortunate that neither L. R. Wager nor J. L. Longland, who had done so well in 1933, was available. Where men have had Himalayan experience selection is not as difficult as it might be; a man who has been high before can be counted upon to go high again, and the more often he has been high and the more recently the better he is likely to go. The number of those who do not acclimatize well to height seems to be very much smaller than those who do, but in taking a man who has never climbed outside the Alps there is the risk that he will not acclimatize; so that if men are available who have proved themselves capable of going high, and are otherwise suitable, then the making of an experiment would not be easily justified. Enough is now known about the technical difficulties of the mountain to show that as well as being able to acclimatize well a man must be a good all-round mountaineer; not necessarily brilliant on rocks or ice, but at home on both and capable of looking after others as well as himself on difficult ground.

There was no mystical value about the number seven; when all are climbers it represents a high margin of safety against casualties. It is possible to hold the view that any number more than one constitutes a large party, but it is more the amount of baggage and porters to carry it, than the number of Europeans, which earns the epithet 'large'. In this sense our party was small, though possibly not so small as it could have been. We hoped that the number seven would form a homogeneous party, small enough to move rapidly, and large enough to split up into two or more parties if necessary.

The names and qualifications of the members were as follows: E. E. Shipton (aged thirty) had climbed in the Alps, Africa (Mounts Kenya, Kilimanjaro and Ruwenzori), and had been six times to the Himalaya, on three occasions leading his own party. On Everest in 1933 he reached 27,500 ft., led the reconnaissance of 1935, and was there again in 1936. In 1937 he took a small party to the Karakoram Himalaya where it explored and mapped a large area of difficult glaciated country in uninhabited regions destitute of supplies. The work done received the approval of the Surveyor-General of India, Brigadier C. G. Lewis, to whose sympathy Himalayan climbers owe much. As Shipton and I had climbed much together our views on the composition and methods of Himalayan expeditions were the same; he was to take my place if the necessity arose.

The record of F. S. Smythe (aged thirty-seven) is well known. He had had great Alpine and Himalayan experience, having been five times to India, first in 1930 with the International expedition to Kangchenjunga, then to Kamet (25,447 ft.) which his party climbed, and in 1933 and 1936 to Everest. In 1937 he had been in Garhwal with Oliver and had climbed seven peaks including the difficult Mana peak (23,860 ft.).

N. E. Odell was in the front rank of mountaineers with long and varied experience in the Alps, the Rockies, the Polar regions, and the Himalaya. His efforts on Everest in 1924 in support of Mallory and Irvine are well known, and it has always been a matter for regret that he did not have a place in one of the climbing parties. In 1936 he climbed Nanda Devi (25,660 ft.) when he seemed so much fitter than the rest of us that I considered his age (forty-seven) to be immaterial. The years between twenty-five and thirty-five have been laid down as the best for high-altitude climbing, but as in most things much depends on the man. The average age of this party worked out at thirty-six.

Dr C. B. M. Warren (aged thirty-two) was a doctor as well as a mountaineer with considerable Alpine and Himalayan experience. He was climbing in Garhwal with Marco Pallis in 1933, and on Everest in 1935 and 1936. He had devoted much time to the study of the use of oxygen and had carried out some practical tests in the Alps, and in 1935 and 1936 he collected physiological data. He was in charge of our oxygen apparatus.

Peter Lloyd (aged thirty) had had much Alpine experience and was first-rate on rock and ice. In 1936 on Nanda Devi he carried a load to the high bivouac at 23,500 ft. proving himself capable of dealing with difficult rock at that height. As a chemist he was able to help Warren with the oxygen.

Capt. P. R. Oliver[1] (aged thirty-two) of Coke's Rifles climbed Trisul (23,400 ft.) in 1933 with one porter. This was the second ascent, the first being the memorable one by Longstaff and the two Brocherels in 1907 when the party climbed from a bivouac at 17,600 ft. to the top and back in one day. In 1936 Oliver was with the Everest party and in 1937 in Garhwal with Smythe. He was therefore another member of the party with the benefit of recent acclimatization.

For my part I had climbed in the Alps, Africa, and the Himalaya, which I first visited in 1934 with Shipton when we explored the Nanda Devi basin. In 1935 I had accompanied Shipton's reconnaissance party to Everest when E. H. L. Wigram[2] and I had degenerated into mere peak-baggers, collecting seventeen all over 20,000 ft. In 1936 I had been with the Anglo-American party to Nanda Devi, which we climbed, and with Shipton again in 1937 to the Karakoram Himalaya. The obtaining of the necessary leave of absence by some members of the party was not easily arranged and in some cases involved sacrifices.

[1] Killed in action in Burma, 1945, commanding a battalion of the 13th Frontier Force Rifles.

[2] Killed, climbing in Wales, 1945.

The personnel having been chosen it remained to collect the material. After the 1936 expedition, Shipton had drawn up estimates of equipment for an attempt on Everest by a small party and to these we adhered closely. The final cost was very near the estimated cost and on the right side, £2,300 against £2,500. The equipment had been ordered in the spring of 1937 before he and I had left for the Karakoram, so that on our return from there in the autumn there remained but the food to be bought and the oxygen question to be decided. The sole innovation in the matter of equipment was that we took only essentials and not too many of those. Three very good types of tents have now been evolved from bitter experience so that many of the discomforts suffered by earlier parties are things of the past. It cannot be pointed out too often that each party draws on the experience and profits by the example of its predecessors—we, of course, benefited most. The types of sleeping-bags, warm clothing, and wind-proof suits, as now used will not easily be improved upon, but the search for the ideal boots and gloves for high climbing still goes on. Thanks to the interest shown by the makers, a Primus stove that burns well at extreme altitude had already been evolved, but this year, at the suggestion of Mr P. J. H. Unna,[1] an improved form of pump-plunger was fitted which made pumping at great heights very much less laborious—a dozen strokes sufficing instead of possibly a hundred. We had to be careful not to fit these altered pumps to the stoves used at lower levels because the extra pressure they gave might have resulted in an explosion.

For the unscientific it might be worth while noting the alteration in atmospheric conditions during an ascent and its consequence:

Atmospheric pressure and consequently barometer readings fall. Air expands as it rises and therefore cools; each cubic foot of air weighs less and therefore contains a smaller weight of oxygen, the proportion between nitrogen and oxygen remaining the same. The quantity of air available for intercepting the sun's rays and for preventing radiation from the earth becomes less thereby speeding up evaporation. The following consequences ensue: warmer clothing becomes necessary not only on account of lower temperatures but also because anoxaemia or oxygen lack reduces vitality. Better tents are needed to reduce radiation; it becomes more difficult to

[1] Mr Unna, in his role of helpful scientist, appears in a more baleful light in the final chapter.

strike matches; to keep candles alight; and for fuel to burn. Cooking becomes more difficult and slower; the lower boiling point and greater cold acting together make it more difficult to keep drinks hot. Strong sunlight may cause snow-blindness and with rapid evaporation may cause sunburn; rapid evaporation causes increased thirst. Reduced external pressure entails a relatively higher internal pressure in enclosed vessels, therefore corked bottles break more easily and cans of liquid tend to leak. Rubber perishes more easily on account of the cold.

All these effects are noticeable in the Alps but are intensified and of more importance on higher mountains.

The question of food does not present any great difficulty until heights of 22,000 ft. or more are reached, if a few simple rules are borne in mind. The technique of travelling light which Shipton and I employ on our own expeditions does not mean that we deliberately starve ourselves or our porters. It does not and should not imply inadequate or indifferent food. As we once lived perforce for a few days on tree mushrooms and bamboo shoots there is a general impression that this is our normal diet, eked out with liberal doses of fresh air, on which, thanks to a yogi-like training, we thrive and expect every one else to do likewise. Nothing could be farther from the truth. Like Dr Johnson, we mind our bellies very strenuously: 'for I look upon it', he said, 'that he who will not mind his belly will scarcely mind anything else.' The more restricted a ration is the more need is there for careful thought in its selection. For normal men a ration of 2 lb. a day is ample (I have kept Sherpas happy for two or three weeks on 1½ lb.), and the whole art lies in getting the most value for weight.

On Polar expeditions where conditions are more severe, the work as hard and the period more prolonged, the sledging ration varies from 25 to 33 oz. and it is usually far simpler and apparently more unappetizing than the food eaten on Himalayan expeditions. The difference is that the men are hungry. Watkins wrote of his 1930 expedition in Greenland: 'We soon found that these rations (39 oz.) were more than enough for men travelling between 20 and 30 miles a day at low temperatures at a height of about 8,000 ft., and towards the end of our time in Greenland we reduced the rations by about one-third, so that the total amount eaten by one man in a day was 23·6 oz. This was found ample for all normal winter sledging work. On all future expeditions I would keep the

1. Upper Sikkim: early snow (p. 34)

2*a*. Gayokang: last halt on Sikkim border (p. 36)

2*b*. On trek to Sebu La: Kangchenjau in background (p. 36)

rations per man per day under 1½ lb.' Martin Lindsay, on the other hand, wrote: 'We found ourselves ravenously hungry on 26 oz. a day.... Were this journey (the crossing of the Greenland ice-cap) to be done again I should increase the ration to 32 oz.' If 2 lb. a day is adequate for that sort of journey it is more than enough for Himalayan expeditions. In 1938 we had food in abundance. The supply of candles may have been short, but we were never reduced to eating them.

On this weight of 2 lb. a day estimates are based, and consequently foods which contain the greatest value for weight should be chosen. Jam, for example, is hardly worth taking; the only useful part of it is the sugar so that a 1 lb. tin of jam, which probably weighs 1 lb. 3 oz. gross, will contain only 10 oz. of actual food. From a quarter to a fifth of the weight of all tinned food is provided by the tin, but perhaps even more powerful arguments against its use are cost, lack of freshness, presence of preservatives and sameness. After a fortnight or so of a diet of tinned food all taste the same—sardines or ham, salmon or spaghetti. Fresh food is always to be preferred to tinned food and as much use as possible should be made of whatever local supplies there may be in order to save transport and money. The proportions of food to be taken can be allocated roughly to 30% protein, 10% fats, and 60% carbohydrates, dividing the last item into half sugar and half cereals. Simple foods are better than processed foods—nobody but the maker knows what has gone into a packet or a tin. Coarse local flour is cheaper and better than white flour, unpolished rice than polished, sugar from a local mill (*jaggery* in India, *gur* in Africa) than white refined sugar. This last is almost as good to eat as home-made fudge, but there is so much moisture in it that the main sugar supply must be white sugar. Of this ½ lb. per man per day is not too much to allow; this 1938 expedition is the only one on which I have not run short of sugar.

Transport to Everest is so simple and is now such a cut-and-dried affair that considerations of weight have not to be so rigorously regarded as in difficult journeys—there is nothing like having to carry or pull your own loads (sledging in the Arctic, for example) for teaching sense in this respect. Consequently we did ourselves pretty well. Our diet included much that was not really necessary;

much which (as the sage in *Rasselas* says of marriage) was rather to be permitted than approved. The principles followed in its selection were simplicity and the avoidance as far as possible of tins. There is no rule without an exception, and in this instance it was the welcome present of a case of tinned tongue; an offer of a case of champagne from the same donor[1] I was short-sighted or hard-hearted enough to refuse and have experienced twinges of regret ever since. It only occurred to me later, when faced like Moses with the murmurings of his people, that the presence of this would in every sense have cut the ground from under the feet of the Sybarites.

The aggregate result of my selection was not so simple as it would have been for a private party but was far too simple for some of us. We took plenty of bacon, ham, cheese, butter and pemmican. Eggs we knew could sometimes be obtained on the march, and by eggs I mean half a dozen or more each—fewer are not much good, but for use on the mountain we had 600 eggs preserved in water-glass, so that every morning we had the Englishman's breakfast fetish of bacon and eggs, even on the North Col. As far up the glacier as Camp III (21,500 ft.) we ate normal food like meat, potatoes, vegetables, rice and lentils; the meat consisted of either freshly killed sheep or yak meat, or failing that dried meat which is a staple commodity in Tibet. For two or three rupees one can buy a whole dried sheep's carcass and this is rather better eaten raw than cooked. That we were able to eat food like this was due mainly to our having a pressure-cooker without which cooking at these heights is almost impossible owing to the low temperatures at which water boils. At 20,000 ft. the temperature of boiling water is only 180° F. instead of 212° F. With the help of dried yeast we made very excellent bread and scones from the coarse local flour—very unlike the white, flaccid, spongy stuff sold[2] as bread in this country, the product of self-raising starch and chemicals. This meagre diet, which might be all very well for ascetics, was eked out with additional items such as milk, porridge, jam, honey, dried fruits, sweets, chocolate, sugar, glucose, dripping, biscuits and soups.

[1] Mr R. W. Lloyd, Hon. Treasurer of the Mount Everest Committee.

[2] 'Which used to be sold' would be more correct though the present article is just as anaemic and less attractive to look at.

Nevertheless, some of us were mightily relieved to find at Rongbuk large quantities of stores left behind by the 1936 party, and considered that this windfall, consisting mostly of nourishing food like jam, pickles, and liver extract (of which there were several cases), alone saved the party from starvation.

It is at heights above 22,000 or 23,000 ft. that the problem what to eat becomes acute, more especially if a week or more has to be spent at such heights. This admission may please the 'caviare and quails in aspic' school of thought, but the fact is that so far it has been found impossible to eat enough of any kind of food in those conditions—even the supposedly tempting foods out of tins or bottles for which the shelves of the high-class grocers have been pretty well ransacked. Owing to its weight the pressure-cooker has to be left behind so that cooking, apart from frying or merely heating things, becomes impossible. One is therefore driven back upon preserved and processed foods out of tins and jars, and the disinclination to eat anything which is already making itself felt thus becomes even stronger. Eating is then a distasteful duty rather than a pleasure, but whether food eaten under such circumstances is of any benefit is a question for physiologists. It is the absence of hunger which makes the problem of Everest so different from that in the Arctic where the sledging ration, within the necessary limits of weight, has only to be designed to maintain the bodily heat and energy of hard-working men. There, concentrated foods rich in fat are the solution which any dietician can work out, and there such foods can be eaten with gusto; but a way of maintaining the heat and energy of equally hard-working men who are not hungry, and to whom the thought of food is nauseating, is less easily found. If you do succeed in getting outside a richly concentrated food like pemmican a great effort of will is required to keep it down—absolute quiescence in a prone position and a little sugar are useful aids. Without wishing to boast I think the feat of eating a large mugful of pemmican soup at 27,200 ft. performed by Lloyd and myself, is unparalleled in the annals of Himalayan climbing and an example of what can be done by dogged greed. For greed consists in eating when you have no desire to eat which is exactly the case anywhere above Camp IV. Of two equal candidates for a place in the party it might pay to take the greediest, forbearing his disgusting idiosyncrasy at low levels for the sake of his capacity to eat at higher. Glucose is good to eat while actually climbing at these great heights, and the quantity of sugar consumed either neat or in tea makes up something of the deficiency suffered in other respects. All the same a straight meal of sugar or glucose hardly gives one the comfortable sensation of having dined. The loss of weight and consequent weakness which follows a stay at high altitudes is probably due as much to lack of food as to the effort expended. I think more use could be made of eggs. They are not very difficult to carry and can be easily cooked in a variety of ways or eaten raw. Sardines, dried bananas, pickled beef, kippers, cream cheese and fruit cake, are some other suggestions, made, be it noted, at Camp III, and not at Camp VI where they would have to be eaten. The fact is that lying in a sleeping-bag doing nothing is, at those heights, so pleasant that some show of resolution is required even to begin the simplest preparations for eating, particularly when these entail the search for food amongst the bedding, spare clothes, boots, rope, stove, candles, cameras, saucepans, and snow littering the tent floor.

Whether to take oxygen or no was an open question which was finally decided in the affirmative for the rather cowardly reason that if we encountered perfect conditions on the last two thousand feet and were brought to a standstill purely through oxygen lack, not only might a great chance have been lost but we should look uncommonly foolish. My present view of the very unsatisfactory nature of even a successful oxygen attempt is recorded in a later chapter, but at that time I mistakenly felt that half a loaf would be better than no bread. Moreover, Warren, who had been trying out a new type of apparatus in the Alps, was very enthusiastic about its possibilities, although I gathered that the opinions of those with whom he had been climbing were more qualified. Some oxygen and some kind of breathing apparatus had to be taken anyhow for medical use in case of frost-bite or pneumonia.

Messrs Siebe Gorman have always taken a great interest in our oxygen apparatus, and this year they spared no pains in fulfilling our requirements at very short notice. These consisted of two of what we called the 'closed' type, and two of the simpler, old-fashioned 'open' type. In the first, to which Warren and most other experts, pinned their faith, pure oxygen is breathed through a mask covering the mouth and nose, the carbon dioxide in the expired breath being absorbed by soda-lime carried in a container fitted above the oxygen cylinder. Although only four hours' supply of oxygen is carried, the whole thing weighs 35 lb., but the smaller supply of oxygen is compensated for, first by the fact that less oxygen is used than in the 'open' type because none is wasted, and secondly it has the effect of providing the wearer with the atmospheric conditions of sea-level, and even gives him a little additional 'kick' into the bargain, so that he should move quicker and thus need less oxygen to complete the climb. We were told of a fireman who, wearing the apparatus, ran for four miles and then carried a man up a ladder, but who failed dismally in attempting the same feat without the apparatus. Inspired by this anecdote I took one of this type to the Lake District and rushed violently up a steep place wearing the apparatus but not carrying a man. From Stockley Bridge up Grain Hill to the Esk Hause path took 45 minutes, and from the path to climb Central Gully on Great End another 25 minutes. Next day over the same track, carrying the apparatus but not wearing the mask, the first part took five minutes less; in the Gully where there was climbing instead of walking the difference was more marked. There was ice and snow in the Gully and the mask hindered while the oxygen did not help. Such tests at anything but high altitudes are of no value, but I felt then that the bad effect of wearing a mask when undergoing great exertion would not easily be obviated.

Professor G. I. Finch, who had had practical experience of using oxygen on Everest itself, held strong views about the unsuitability of this type and the simplicity, lightness and greater comfort of the 'open' type. At his urgent advice I had two of these made. They carry half as much oxygen again, weigh 10 lb. less, and there is no mask; the oxygen is supplied through a tube to the mouth, and air, such as it is at that height, is breathed in through the nose. The difference in weight is accounted for by the absence of soda-lime and a simpler mechanism, which is of course, less liable to go wrong.

It was with a similar type that Finch's party made their great effort in 1922 when they reached the highest point then attained—27,230 ft.—which was about 200 ft. higher and nearer the summit in lateral distance than the non-oxygen party reached. Moreover, Finch's party was not well acclimatized, having spent only five days no higher than 21,000 ft.; it was a weak party, for Finch himself was the only experienced mountaineer; and judging from the times given they appear to have been slightly the faster party of the two. Both parties suffered slight frost-bite and both on coming down appeared equally exhausted. There is no doubt that oxygen will greatly assist those who wish to go high before they are properly acclimatized, a process which involves time and trouble. Indeed, it is arguable that the benefit of acclimatization attained by living at about 23,000 ft. for a few days is more than offset by the loss in strength, and it was Finch's idea that the mountain should be 'rushed' by an oxygen party which had not had its strength sapped away by too much acclimatization. But even this 1922 evidence, which is the best we have, is inconclusive. The party was moving on easy ground where it was not considered necessary to rope, and higher up, where the climbing is far more difficult, the oxygen apparatus might be more of a hindrance than a help. Of Mallory's experience with it in 1924 we shall never know, while Odell in the same year found he went as well without as with oxygen.

Before shifting the scene to India I must deal shortly with the financial arrangements. We budgeted for a cost of something like a quarter of previous expeditions—£2,500[1] as against £10,000—and more than enough to cover the estimate was subscribed by generous friends and by members of the party themselves. I asked members to do this because I thought that if each of us had a small financial interest there would be more incentive to economize. It also seems to me to be the proper thing; an expedition to Mount Everest may not be every one's idea of a climbing holiday, for which no one could object to paying something, but it is a great privilege for which many would sacrifice more than money.

We had sufficient money to do without newspaper support, and at first we thought of doing without because there was general agreement amongst mountaineers that publicity concerning Everest had increased, was increasing, and ought to be diminished. But some of our subscribers quite rightly hoped to be paid back; and since the mountain is now a matter of world-wide interest it is

[1] Actual amount spent £2,360.

almost essential to have some official news channel, for otherwise the Press in general will and does see to it that there is no lack of unofficial news. We made arrangements with *The Times*, who treated us generously. This was satisfactory from our point of view in that there was no obligation upon the leader to send back long messages while actually at work upon the mountain, but it was a matter of regret on our part that through an unfortunate incident *The Times* had good reason to complain of our treatment of them. In the forlorn hope of damping down our news value we intended to send as little news back as possible and then only important news. What was sent, therefore, was scanty, but sufficient to keep informed those who prefer a straightforward statement to surmise, sensation, and ballyhoo.

CHAPTER III

PREPARATIONS IN INDIA AND DEPARTURE

We're clear o' the pine and the oak-scrub,
We're out on the rocks an' the snow.... KIPLING

SHIPTON and I reached Kalimpong on 14 February, a fortnight in advance of the rest of the party, where we were very kindly received by Mr and Mrs Odling. Two old friends, Angtharkay and Kusang Namgyal, were already installed there undergoing a course of instruction in cookery—the long white aprons they affected gave them a most comical air. Not that Angtharkay wanted very much instruction in the sort of cooking we should expect of him; his stews and curries had always been masterpieces, and, like most natives of India, he could, of course, turn out rice to perfection. To students of recent Himalayan literature his character must be already familiar. This year he added the duties of cooking for seven of us to his usual jobs of bossing up the porters, bestowing valuable advice and encouragement upon his employers, and carrying a heavier load farther and higher than any one else if called upon to do so; in fact, a sort of Jeeves, Admirable Crichton, and Napoleon rolled into one, but taking himself less seriously than any of those redoubtable men. Kusang I had not seen for four years, but had no difficulty in recognizing him in spite of the apron and the fact that he was not crooning the mournful dirge of two notes and three words which I shall always associate with him. He seemed to be less carefree than in those far-off days; I think he had married. We had to take him out of his apron next day to send him off to buy food in Sola Khombu, and to arrange for it to be brought to Rongbuk by thirty Sherpa porters on 7 April or thereabouts, all of which he did with great faithfulness.

Karma Paul, the zealous and energetic interpreter and general factotum, whose sixth Everest expedition this was, had already met us at railhead on the way to Kalimpong and had taken charge of us. Everything seemed to be either done or in hand except the actual climbing of the mountain. Next day he and I went over to Darjeeling to select porters, armed with a list drawn up by Shipton and myself, or perhaps more correctly, by Angtharkay. No one who was not vouched for by him stood very much chance of being selected. This was not so unfair as it may sound. There was no nepotism, but all were either friends of Angtharkay or their characters were well known to him. As Lord Fisher used to say: 'Favouritism is the secret of efficiency'; and there is some justification for confidence in the friends of a man who sets such a high standard for himself. Nor is it likely that a man would stretch a point in favour of a friend whose failure would be likely to throw more work upon himself.

Besides picking twelve porters for ourselves we had to choose twenty for three other parties coming to the Himalaya later. After this was done there seemed to be no good experienced men left. One, Nursang, who had accompanied several German expeditions as head-man, was collecting porters for Nanga Parbat, and was having some difficulty in persuading men with the memory of so many dead comrades fresh in their minds, to go again to that unlucky mountain. Nursang's forcible appeals reminded one of the exhortation of Frederick the Great to his wavering troops: 'Come on you unmentionable offscourings of scoundrels. Do you want to live for ever?'

The old hands amongst our lot were Pasang Bhotia with whom I had travelled in 1934, when with Angtharkay and Kusang Namgyal he was the third of an incomparable trio; and also in 1935, the occasion when he had combined business with pleasure by marrying a good-looking buxom Tibetan girl on the way back from Everest —the girl has since died, but the fate in store for poor Pasang himself was even more unkind.[1] Then there was Rinsing, who is a grand chap once he gets above, not the snow-line, but the beer-line—that is beyond the reach of villages and liquid temptation; Lhakpa Tsering, who has the manners and appearance of an Apache, but does more work than several others put together; Tensing, young, keen, strong, and very likeable; and Nukku,[2] bovine but tireless. At Smythe's earnest request we also took Ongdi Nurbu

[1] He was struck with paralysis on returning from Camp VI, and although he recovered the use of his limbs some weeks later, he would probably never be able to climb again.

[2] This was the Nukku who died of cerebral malaria when with me in the Assam Himalaya the following year.

and Nurbu Bhotia. Ongdi is in a class only with Angtharkay; of a character that makes him the natural leader of the other porters who delight to call him Aschang or Uncle. Besides being a tower of strength in any caravan he is an exceptionally capable mountaineer, but is handicapped by a susceptibility to pneumonia which he has now had twice on Everest; but if he contracts it easily he manages to throw it off with still greater ease.

A telegram from the railway which read: 'Expedition bits dispatched by forty up train', advised us of the imminent arrival of our stores and not that which it might reasonably have led us to hope or expect. The young mountain of bales and boxes of all shapes and sizes which was presently to be seen covering the floor of a godown, made for me a depressing sight—more depressing still perhaps had one been a Sikkim mule or a Tibetan donkey—but with the help of Karma Paul and the Sherpas it was soon divided into loads of more or less suitable size and weight. I had brought out some particularly strong gunny bags which took the bulk of the stuff and stood the journey sufficiently well. The lightest box that will hold 80 lb., itself weighs about 10 lb., so that for every eight useful loads there is one additional load of almost useless boxwood. A box is just as easily stolen bodily as a bag (Odell will confirm this) and not much more difficult to break into by a determined thief. There was only one attempt at pilfering on the march; and the only serious loss on the way up occurred when we were beyond Rongbuk, when there were with us in camp only Sherpas.

On the march back a box of fossils and stones of scientific interest, highly prized by Odell, was taken from outside his tent where it was playing the part of anchor for one of the guy ropes. The thief no doubt mistook it for coin (the box was very heavy) and whatever amusement one felt at the thought of the chagrin awaiting the thief, had to be severely repressed out of respect for Odell's cold fury. It was a serious matter any way one looked at it. The results of several months' work had gone, and if the thief talked, as he might, for sorrows shared are sorrows halved, the Tibetan officials might learn we had been collecting stones against their express orders. They quite rightly object to any one knocking their rocks to bits with a hammer, thus releasing any malign spirits which may happen to be in them. We could only pray that the thief's professional pride would restrain him from advertising the fact that he had risked his life or the skin of his back for a box of pebbles.

It seemed that fifty mules would be sufficient to carry the stuff, but with the arrival of the oxygen, most inconveniently packed, the number rose to fifty-five and later had to be increased. The transport of the oxygen outfit is always a source of worry to those concerned, but perhaps no worse than that of the kerosene oil. The cylinders have to be closely watched for leaks, the tins of soda-lime have to be protected in boxes, and the apparatus itself, whether packed in a vast coffin shored round with sweaters and socks, or carried on the back of a trustworthy porter, always seems to develop some defect. Suitable containers for kerosene are not easily found:[1] I spent much time and cracked one of Odell's geological hammers in keeping the leakage of ours within reasonable bounds. After every march screw-tops had to be tightened up or leaks soldered.

Meantime, the Sherpas had to be sent to the hospital for treatment for worms. One was found to have boils and had to be left behind altogether, and Rinsing and Pasang developed some other complaint—I think the local beer must have been bad—which necessitated their following us later with Smythe. A second visit had to be made to Darjeeling to draw money (3,000 silver rupees and 5,000 in paper); oil drums had to be tested for leaks and mostly rejected; sugar (880 lb.), rice, atta, vegetables, candles and matches, presents for Tibetans, cooking and eating utensils for ourselves and the Sherpas, had all to be purchased. Regardless of expense, every European was given a plate in addition to a mug, and there were even dish-cloths for wiping them. Altogether there was sufficient to do, especially as Shipton had retired into confinement for the speedier delivery of a book which he had carelessly omitted to finish before leaving home.

The rest of the party, except Smythe, who followed later, arrived on the last day of the month and were immediately handed a little list which Shipton and I had thoughtfully drawn up for guidance in deciding what to bring. I have no doubt it was read with the tolerant interest bestowed on such things. Although the mules were to leave before dawn next day all was ready in time—

[1] The 'jerrycan' will solve this problem.

even Odell, whose baggage I received in the godown at 8 o'clock that night. It was a pleasure to see he had not forgotten his glacier drill.[1]

The mules were not quite the first of us to get away. Two days earlier I had sent off a very redoubtable advance courier in the form of Purba Tensing. He was to act as mail runner for us, as he had done before, but meantime he had gone to Shekar Dzong armed with a wad of rupee notes to buy *tsampa*. Besides many others, Purba has three qualifications which Montaigne considered necessary for a good servant: 'That he should be faithful, ugly and fierce.' He is an Eastern Tibetan; a big man with an air of amiable ferocity, carries a long Tibetan knife, walks or rides great distances with equal facility, and is a whole-hearted believer in small light expeditions. He travels alone and carries nothing at all. The last glimpse I had of him was cantering swiftly through the Kalimpong bazaar bound for Shekar 200 miles away, and if he was carrying any luggage it must have been concealed about his person. We met again at Guru, in Tibet, three weeks later, when he very apologetically asked me for some more money to defray travelling expenses. To show that he was exercising due economy and that the claim was not unwarranted or inflated, as claims for travelling expenses sometimes are, he added that on the last march to Guru from Shekar he had been obliged to boil his shirt in order to make a cup of tea. An infusion of shirt might look like tea—might even taste like it—for Tibetan tea is not as ours.

The mules were to reach Gangtok on 3 March where the loads would be taken over by Lachen muleteers who were to accompany us as far as the first march into Tibet. Last-minute purchases and presents of vegetables from well-meaning friends had resulted in raising our mule requirements to sixty—a number which almost justified the question of the reporter of an Indian newspaper: 'When does the expeditionary force start?' It suggested, too, that a little more ruthless 'scrapping and bagging' would not have been amiss. Another question which I thought pleasing, addressed to me by an otherwise intelligent man, was: 'And is Mr Smythe a climber, too?'

We ourselves left by car on 3 March to reach Gangtok at 2 o'clock that afternoon. The road was very greasy after a night

[1] Vide *The Ascent of Nanda Devi*, chap. XII.

3*a*. The route to Shekar Dzong (p. 43)

3*b*. View of plains from battlements of Shekar Dzong castle (p. 44)

4. The fortress of Kampa Dzong with the Tibetan plains and the peaks of Northern Sikkim (p. 39)

of heavy rain. Even at this early stage the weather was arousing comment; owing to heavy clouds we had scarcely seen the mountains during our fortnight at Kalimpong. Mr B. J. Gould,[1] the Political Officer, who had already shown us many kindnesses—amongst others he had made our Sikkim transport arrangements—added to our indebtedness by having the Residency opened for us in his absence. Here the Maharajah's sister, the wife of the Bhutanese Prime Minister, very charmingly acted as hostess in Mr Gould's absence. There was much to be done and little time to do it in, for that night at the Palace we had to attend a dinner which H.H. the Maharajah was very kindly giving for us. The loads had to be checked; kerosene tins, already leaking, soldered or re-washered; the Gangtok mule-men to be paid off and the Lachen men given an advance; all before a perfunctory wash—the last—and a battle with a boiled shirt. Some careful planning had to be done in the matter of dinner kit. As we did not want to dress every night on the mountain, or even on the march, we packed our things next morning in two communal suit-cases and returned them to Kalimpong. There was one slight hitch which only became manifest next August when I was dressing for dinner at Government House in Shillong, when I discovered that my shoes were about three sizes too small. The mountaineer in India, particularly the member of a large expedition, must firmly suppress any latent tendency he may have to go native or even to slovenliness. Once in Darjeeling I was obliged to borrow a suit of Karma Paul's as my own was condemned as unsuitable for a luncheon party; and on another occasion to wear my host's evening clothes which hung about me like a 'giant's robe upon a dwarfish thief'.

Most of the European residents of Gangtok had been invited by H.H. so that it was quite a large party which sat down to dinner that night. Speeches were made, and to me fell the task of replying on behalf of the others to the good wishes expressed by our host. Observing how faithfully they were dealing with the magnificent dinner provided, I remarked that my comrades had all the appearance of men who had either just undergone a long fast or were about to undergo one—a thought which had possibly occurred to others besides myself; and in view of the slightly less luxurious

[1] Now Sir Basil Gould, C.M.G., C.I.E.

diet in store I ventured to remind them of what I thought was a not inappropriate saying of Thoreau, that great apostle of the simple life who practised in the American woods what he preached in his book, *Walden*: 'Most of the luxuries and many of the so-called comforts of life, are not only not indispensable, but positive hindrances to the *elevation* of mankind.'

Our sixty mules left at 9.30 o'clock next morning, but our departure was postponed to a more civilized hour, namely after lunch, which enabled us to complete our postal arrangements and write final letters. The Maharajah and a large gathering gave us a hearty send-off from the Residency, but before we were allowed to depart much film was wasted on an uncommonly sheepish-looking group doing its best to pass for intrepid mountaineers. Even at this eleventh hour kind friends interposed between us and too sudden an introduction to hardship by carrying us in cars to the extreme limit of the road just below the Penlong La. From here the remaining eight miles to Dikchu is all downhill so that we were able to complete the first stage of a 300-mile journey without unduly exhausting ourselves.

On Everest expeditions an unfortunate tradition has grown up that the Sherpas carry nothing until the mountain is reached. It is in strict accord with the principle of the conservation of energy, but I cannot think that the carrying of 25 or 30 lb. for three weeks would have any seriously debilitating effect upon men who habitually carry twice as much for as many months. This time I gave the twelve Sherpas a mule between them to carry some food like rice and lentils, which are not obtainable in Tibet, and cooking gear, but their own kits they were expected to carry themselves. It was a pleasant surprise, therefore, when I found them insisting on carrying our light rucksacks containing odds and ends; I thought it showed praiseworthy keenness until I found they had distributed their own kits amongst the mules in order to carry our sacks which were very much lighter. We had encountered the same trouble in 1935, when an effort to save transport costs by making the Sherpas carry loads resulted in a strike. The tradition has now spread beyond the confines of an Everest expedition, and it is wise to discount the Sherpas as a carrying force so long as any other transport is available. They either put their loads on the already sufficiently laden animals or hire animals on their own account and present you with the bill. The evil of large expeditions is thus made acutely manifest to those who follow; and not only in this way, but because they set standards of work and pay which are not always fair to the employer, and which generally spoil the market for goods and services for all time. Any one who goes to Askole, for instance, in the Karakoram, and offers the usual rate of Rs. 1 per day is regarded as either a skinflint or a 'poor white' because a large expedition which went there before the war paid their porters Rs. 2. We ourselves, at a place in Tibet, were regarded as cads, or at any rate, not quite gentlemen, because the tip we gave to the servants of a house in which we had been entertained was thought small; as indeed it was in comparison with what had been given in the same place on previous occasions.

Over the first few days of any march it is wise to draw a veil. The things that have been forgotten are gradually remembered, and the whole organization creaks and groans like your own joints. You wonder if man was really intended to walk, whether motoring after all is not his natural mode of progression, and whether the call of the open road is as insistent as you yourself thought or as the poets of that school sing. Nothing much happened. It rained; but the Sikkim marches are mercifully so short that generally we managed to avoid it. At Mangen, at what corresponds to the local coffee-stall, we had the usual 'elevenses' of sweet tea in a glass, soup, curried eggs, and pork pies made from the lean hyena-like pigs rooting about in the dirt outside. At Singhik we drank *marwa*, the local beer, with our supper. Six great bamboo jorums about a foot high, in which the beer is served, give the table a falsely convivial air, for the contents are 99% millet and 1% beer which is laboriously sucked up warm through a bamboo tube. I found a frog on the table that evening and was curious to know from whose beer-jug it had come.

On the march before Lachen a landslip temporarily dislocated the traffic. It was a small slip but the smallest is serious where the narrow mule track is carved out of the face of a precipice. After some repairs had been done we got the whole mule train safely over, but just beyond, one of the mules carrying fodder, shied and stepped off the path, to be instantly killed in a sheer fall of 200 ft.

to the Tista river below. The owner, a poor woman, who had to be compensated for the loss, was most distressed. We were very relieved to know that none of our baggage was lost, but we persuaded Tensing, Warren's servant, to break the tragic news to his master that the mule carrying all his kit had gone over the 'khud'. Wretchedly bad actor though he was, Tensing managed to convince his master that the worst had happened; but we were disappointed of our jest because Warren, instead of acting in the expected fashion, beating his head against a stone or tearing his hair, expressed no more concern than he might have done at the loss of a pocket handkerchief.

At Lachen we heard rumours that the road was blocked with snow. Inquiries about the state of the road had of course been made before we left but no definite information had been forthcoming. Leaving the party to rest after its labours, Shipton and I took horses and rode up towards Tangu some 13 miles away. Within half a mile of Lachen snowdrifts were met, and six miles this side of Tangu the snow lay everywhere two feet deep. A narrow track had been shovelled up which we rode until we met a gang of forty men at work on it. These had been sent up by the Sikkim Road engineer in accordance with the wishes of the Maharajah that we should receive every assistance. The men expected to clear a track through to the Tangu bungalow by next day but they told us that snow lay deep for another 10 miles beyond. It would evidently be a long job, so next day Lloyd and I went up to the bungalow to superintend the work while the rest of the party remained at Lachen.

Tangu has earned an evil reputation by reason of the exaggerated effect its moderate altitude of 12,000 ft. has on most people, not even excepting seasoned mountaineers. On our arrival I was suffering from a splitting headache and Lloyd laid all before him; nor were we very much better the following morning, Lloyd losing his breakfast. The others, when they arrived five days later, were less troubled, but Smythe told me that when he came through later he too vomited. I have not found any other place of similar altitude in the Himalaya which has this effect upon those coming up to it for the first time. A suggestion has been made that it is due to its great humidity.

Accompanied by the road foreman we walked up the road next day for six miles which seemed like sixteen. There was a couple of feet of snow almost the whole way and for several miles beyond, but on the way home when we met the road-gang already two miles out from Tangu, we thought the job would soon be finished. In spite of their primitive implements they shifted a surprising amount of snow. A fresh fall of snow next day tempered our optimism, and on the 12th, the day the mules had been ordered up, snow began falling at 7 o'clock and continued most of the day. I hoped the mules would turn back, but at 4 o'clock they duly arrived, their loads covered with snow. There was nothing for it but to off-load and send them back to Lachen; for now there was no prospect of the track being open for some days and there was nothing here for them to eat.

It was exasperating for the men to have all their work to do over again, snow and wind having completely filled in the trench they had dug. During the next few days this happened repeatedly. The Tangu bungalow is comfortable enough so that we might have sat there waiting patiently for better weather were it not that we had engaged to pay the mule-men half rates for every day we were delayed. We amused ourselves exhorting the road-men, walking up the nearby hills, and stalking a herd of bharal. The mule contractor lent us our only weapon—an old, cross-eyed carbine which needed humouring—and with it some twenty rounds of ammunition. Oliver, representing the Army, fired seven ineffective shots; a feat which evoked murmurs of 'Thank God, we have a Navy'. Shipton also made no impression on the long-suffering herd and it was left to Angtharkay to bring one down with the last remaining round.

On the 14th the road-men were driven down early by a blizzard. This was bad enough, but they had also finished all the food they had brought up, so in desperation we ordered up the mules for the 16th and persuaded twenty of the road-gang to stay and see the mules through to the next camp, Gayokang. The morning of that day dawned bright after a windy night. Fine weather seemed to have set in at last and the men were much cheered by the sight of birds flying north—tiny white wedges high up against the cold blue sky. In the morning some of us climbed a hill of about 16,000 ft. whence we got a magnificent view of Kangchenjunga and Tent

Peak. The reckless rapidity of our descent by means of sittin glissades over a mixture of snow, grass, and stones, considerabl astonished the natives.

The mules arrived that evening and at last on the 17th we mad a start. The pine forest which at Lachen takes the place of th tropical rain forest, comes in its turn to an end at Tangu. Silve birch, rhododendrons, and juniper bushes, struggle on for a bi higher up the road until these too cease, the grass withers, an soon only a few thinly scattered hardy Alpines are left to fight i out with the stone and gravel wastes. The track was in fair condi tion, but at the eighth mile we encountered an undug drift whic delayed us for half an hour. Here we dismissed the road-men, fo as the road climbed the snow gradually grew less until at Gayokan the only remaining traces of winter were some large sheets of ice The reason for this seeming paradox is that north of Tangu th climate becomes more Tibetan in character. The north-west win which blows in Tibet for the greater part of the year drives bac the moisture-laden winds from the south. Little snow falls, an wind and evaporation between them account for its rapid dis appearance. One wondered whence the snow-clad peaks right an left of our bare, brown valley—Kangchenjau, Chomiomo, Lachsi— received all their snow.

The mule bells jangle merrily as the beasts hurry through th stony defile which they well know marks the approach to Gayokang and the end of one day's journey. Round the corner are the smoky hovels; mean and dirty enough in truth, but bright and friendly when viewed against the cold and desolate slopes sweeping up to the northern skyline. Men and beasts crowd in. The mules are picketed under the walls, kicking and squealing, while the men run down the lines putting on the nose-bags containing a woefully smal feed of barley. As dusk falls the men disappear into the flat-roofed huts to cook their meal over the yak-dung fires. With the night comes wind and snow.

5. Kampa Dzong: Gyangkar Range in distance (p. 40)

6a. Collecting yaks for the day's march (p. 42)

6b. Ploughing in Tibet (p. 42)

CHAPTER IV

THE MARCH TO RONGBUK

For if everyone were warm and well fed, we should lose the satisfaction of admiring the fortitude with which certain conditions of men bear cold and hunger. MR PECKSNIFF

AN efficient mule train is one of the most satisfactory means of transport. Mules travel as fast as a man can walk so there is no need for the baggage to leave hours in advance of its owners in order to complete the march in reasonable time. They carry their 160 lb. effortlessly; and knowing what is required of them seem intent on completing the day's stage with as little fuss and delay as possible; whereas donkeys, bullocks, or yaks, wander about snatching mouthfuls of grass, stop, and even lie down, unless constantly driven. Moreover, mules have not that air of patient suffering which yaks, ponies and especially donkeys, assume when travelling under a load—an air which, if the loads happen to be mine, awakens in me a feeling of acute discomfort. A man under a load never arouses this feeling of pity, because men, unlike animals, will not and cannot be driven to carry too much or too far. If they do it at all they do it voluntarily, and stop long before that state of exhaustion is reached which sometimes causes animals to drop dead in their tracks.

These Lachen mules of ours were well up to their job—fine, upstanding animals with clean legs, neat feet, smooth shiny coats, and a general air of breeding. They come from the northern part of Tibet. In Southern Tibet the donkeys are little bigger than St Bernard dogs; so small that no pony could very well mate with them, and so melancholy is their appearance and so hard their lot that it is difficult to imagine they should ever wish to procreate their species. The mule-men, too, on the trade routes between India and Tibet, understand their job and take a pride in their animals which they deck with collars of bells adorned with bright scarlet tassels of yak hair, or scarlet wool tufts, and decorative brow-bands. The drivers wear homespun natural coloured trousers, stuffed into high, red felt boots secured at the knee with brightly coloured garters, and an easy-fitting coat of the same material reaching to the knees. The sleeves of this are long and loose, but the right arm is nearly always kept out of the sleeve which therefore hangs empty. The coat is secured round the waist so that the upper part forms a capacious sort of kangaroo's pouch in which are stored pipe, tobacco, snuff-box, purse, a piece of dried meat, and probably some *tsampa* in a skin bag. A knife, which more often resembles a short sword, is slung at the waist, and a short-handled whip carried in the right hand or stuck into the top of the boot, completes the outfit.

We did not stir until 9 o'clock when the sun had warmed men and mules after a night of bitter cold. After some of the mules had come down with prodigious thumps they picked their way more carefully round the sheets of ice which partly covered the flats at the foot of the long final rise to the Sebu La. As we mounted the bare brown slopes, the wind gained force until on the top (*c.* 17,000 ft.) it was blowing with the peculiarly cold, unrelenting ferocity of Tibetan wind. The sun had long disappeared behind the scurrying clouds when we passed the cairns marking the summit and began the long, easy descent with bowed heads and averted faces. This, I thought, is the kind of weather we must expect in Tibet so early in the year, and the fact that I was wearing shorts lent poignancy to the thought. As we hurried down to Guru, intent on finding shelter, we passed large herds of yaks and sheep grazing happily on gravel, contentedly oblivious of any wind, while the herdsman stood about equally regardless of the piercing blast. The inhabitants of Tibet are tough. For our part, not being so tough, we sought food and warmth in the dark interior of the headman's house, pitching our tents in the lee of the compound wall.

As evening drew on the absence of Odell began to cause uneasiness to those who had not travelled with him before, and who were consequently unaware of the long detours and longer halts which were sometimes imposed on him by his devotion to science. Accustomed though I was to his vagaries, and perhaps not as sympathetic as I should have been towards their cause, the alarm of the others infected even me. Search parties were about to set out when the wanderer cast up. Instead of telling us, as well he might, of the discovery of some intensely interesting rocks which had necessitated prolonged examination, he frankly confessed to having lost his way. Such frankness was worthy of belief, but the Sherpas were sceptical, and unkindly suggested that the afternoon had been spent at a neighbouring nunnery or *ani-gompa*, as they are called. Henceforth he was known to them as the 'Gompa La Sahib', for it was in the direction of that place he had strayed. When, as not infrequently happened, he again took the wrong path, their advice to him about the right one was always accompanied by the assurance that there were no nunneries in that direction.

After this rude welcome into Tibet, next day's march to Kampa Dzong was done in wonderfully clear, calm, cloudless weather. Close on our left Chomiomo and many other peaks sparkled in the sun; farther lay Kangchenjunga and his satellites; while a hundred miles to our left front Everest and Makalu drew all eyes. Nor was this benignant day to be the one exception. Our gloomy forebodings about the rough passage we were likely to have on the way to Rongbuk were falsified by a more or less unbroken succession of sunny days. The wind seldom became violent before noon, but after that it was unpleasant. At Kampa Dzong a furious blast drove the pole through the apices of two bell tents, an occurrence which the Sherpas considered extremely funny. The local weather experts assured us that these comparatively genial conditions were unusual for March, but Shipton thought they were not very different from those experienced in 1933. As in our case, the 1922 and 1933 expeditions encountered bad weather on first crossing into Tibet, but thereafter enjoyed fine but cold weather. In 1924 and 1936 conditions seem to have been definitely milder. General Bruce, comparing 1924 with 1922 speaks of an 'infinitely milder climate', and Mr Ruttledge in 1936 calls the weather 'comparatively warm'; but no useful conclusions can be drawn from this because in the first case the monsoon was later than usual and in the other much earlier.

Two days were spent at Kampa Dzong, the last resting-place of Dr Kellas who died there in 1921. A very remarkable sight is the old fort imposingly situated on the summit of a high yellow rock which on two sides falls away sheer. On another side a long fortified wall runs from the fort to the foot of the rock to give the garrison access to a well. This wall runs across some slabs which in angle and smoothness are not unlike the rocks on the upper part of

Everest, though less steep. On these we scrambled about, hopefully wearing the oxygen apparatus whose weight made it difficult to walk up the easy slope with any confidence. Rope tied round the boots to cover the nails helped. Smythe had brought out a specially made pair of *scarpetti* which he hoped to use on the upper rocks of the mountain; but, alas, these devices we were never to try.

Three memorable things at Kampa were a dinner of blood and rice sausages from a freshly killed sheep, long arguments about the zodiacal light, of which at night there were most brilliant displays, and a highly successful stratagem on the part of the Sherpas. With very long faces they complained that they had eaten all the food brought with them, that owing to bad crops food in Tibet was very dear, and that a special food allowance of two annas a day until Rongbuk was reached, and an immediate advance of Rs. 3 a man, were an imperative necessity. I fell in with this very plausible request only to find that its immediate result was an uproariously successful 'blind'.

When we started again on the 22nd it was amidst the usual indescribable confusion following upon a change of transport. A scratch assortment of ponies, bullocks, zos (a cross between a cow and a yak), and donkeys, took the place of the efficient mule team. There was not a single yak, because, we were told, an epidemic the previous year had reduced the yak population of the district from seventy to seven. There were almost as many owners as animals and of course, each fought strenuously for the lightest and handiest loads for his own animal, and to avoid being left with the heavier and more unwieldy loads. With energetic voice and gesture Karma Paul gradually produced order out of chaos and at length the whole caravan had straggled off with the exception of one donkey which very wisely refused to carry any load at all. For this the headman's own pony had to be impressed.

We all walked on the march up; not because we liked walking but because it was warmer and in some ways less tiring. A good Tibetan pony can amble fast and comfortably for a long time but only if the rider has mastered the correct aids which are by no means the same as those used in the *haute école*. A bad Tibetan pony is worse than no pony at all, as, in addition to the primary discomfort, there is the added danger of a fall; for one sits perched high over the animal's head (they are all very short in the rein), on top of a doubled sleeping-bag, tied insecurely to the saddle, which in turn is girthed still more insecurely to the horse. The stirrups are seldom wide enough to admit a European boot, and the leathers are so short that you have to ride like Tod Sloan, which is particularly fatiguing for long periods at walking pace. The art lies in balance, and though a climber can balance on his feet it does not follow that he can balance on his seat.

At Tengkye Dzong another two days were spent camped by a placid lake whose waters lapped the walls and almost surrounded the village. Even at this early stage Oliver had a cold and Warren a sore throat. The Dzongpen was very friendly and entertained us to several meals of macaroni, stew, radishes, so-called mushrooms, which were really fresh-water weed, and Chinese sauce. Later he tried to teach us Chinese dominoes—a very difficult game which our poor brains, even had they not been dulled by bowls of macaroni and beakers of local brandy, would have had trouble in mastering. Next day we received him officially, a ceremony for which Karma Paul insisted on our washing, and for which Angtharkay got himself up like a bird of paradise in sky-blue coat and scarlet trousers. We produced our passport and some presents before settling down to a long haggle over the rates to be paid for the fresh animals which were to carry for us from here. Owing to rigid custom it is impossible to have one lot of animals to carry right through to Rongbuk; changes have to be made in passing from the jurisdiction of one headman to that of another. On this occasion the Church, represented by the genial but shrewd-looking abbot of a nearby monastery, took a leading part in the negotiations and drove an unconscionably hard bargain.

As the next march from Tengkye is a long one, involving the crossing of two high passes (*c.* 16,000 ft.), our new transport did not get in until dark. On the way up to the pass we caught magnificent views of the Tsomo Tretung lake, its deep blue expanse ringed with warm yellow hills. Most of the Europeans usually arrived in camp long before the slow-moving transport and passed the time until its arrival lamenting its slowness and drowning dull care in Tibetan tea or beer. A few Sherpas were detailed to accompany the baggage and I usually stayed behind myself, more as a matter of form than for anything I could do in the way of hastening them or preventing pilfering. We lost a pair of boots, and one night an unsuccessful attempt was made to remove a load bodily, but that was all. The animals were never off-loaded during a march; the men would sit by the roadside to drink 'chang', which they carried in a skin, from wooden bowls, while the beasts picked up a few mouthfuls of gravel and weeds. At night if there was no gravel and weed but only gravel, the animals would be given a few wisps of wiry hay (imported, one presumes, from India), the yaks receiving a ball of *tsampa* about the size of a child's football. *Tsampa*, of course, was also the men's supper, which simplified the cooking, and having eaten it they stretched themselves luxuriously on their spacious bed—the ground—alongside an extinct fire (fuel is scarce), sheltered by the wide canopy of the sky. There is a Tibetan proverb which says that he who knows how to go about it can live comfortably even in hell, and certainly the inhabitants of that country, human or animal, will not fail in that respect for want of practice.

Following down the valley of the Chiblung Chu we reached Jikyop, or 'The Place of Fear'. It lies in a narrow gorge flanked by steep sand-hills, and is said to have been a favourite place for bandits until the little fort, which is its only building, was built. No wind-tunnel constructed by man could more efficiently produce concentrated gales than this shallow gorge. We left the redoubtable Purba Tensing here, for it was the half-way point at which he would hand over and receive mail from the two men who were to carry it to and from Sikkim. He was content to work alone, but the Sikkim men had insisted on working in pairs.

On emerging from the gorge the track fords a shallow river, which was frozen up, and then crosses a dreary waste of sand dunes before turning up the comparatively green valley of the Phung Chu. By the first river is a hot spring, so hot that it was five minutes by the watch before I was able to get in up to the neck. Numerous little worms, however, swimming about in it, seemed to find the heat very tolerable. After being thus parboiled we found the crossing of the thin ice of the river in bare feet an interesting contrast. The transport animals, especially the poor little donkeys with their small feet, made heavy weather of it in the soft sand; two of them foundered and one of them died. Trangso Chumbab, where

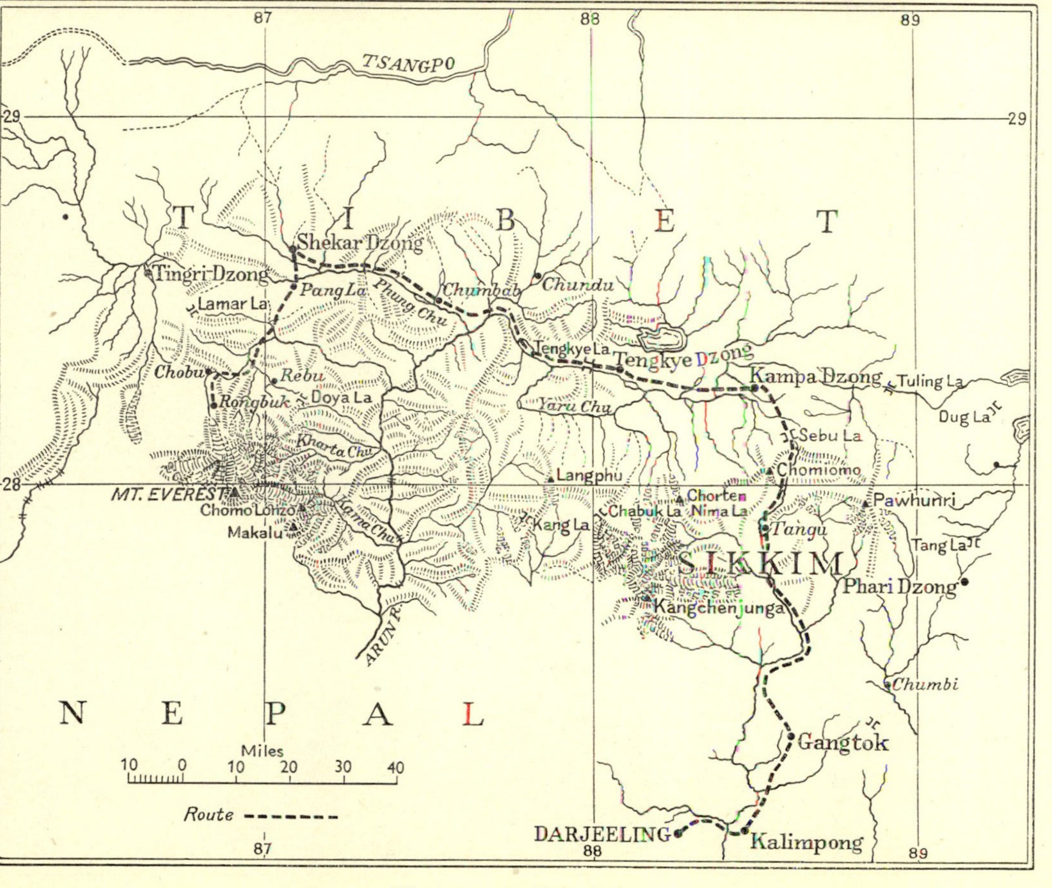

The route to Rongbuk

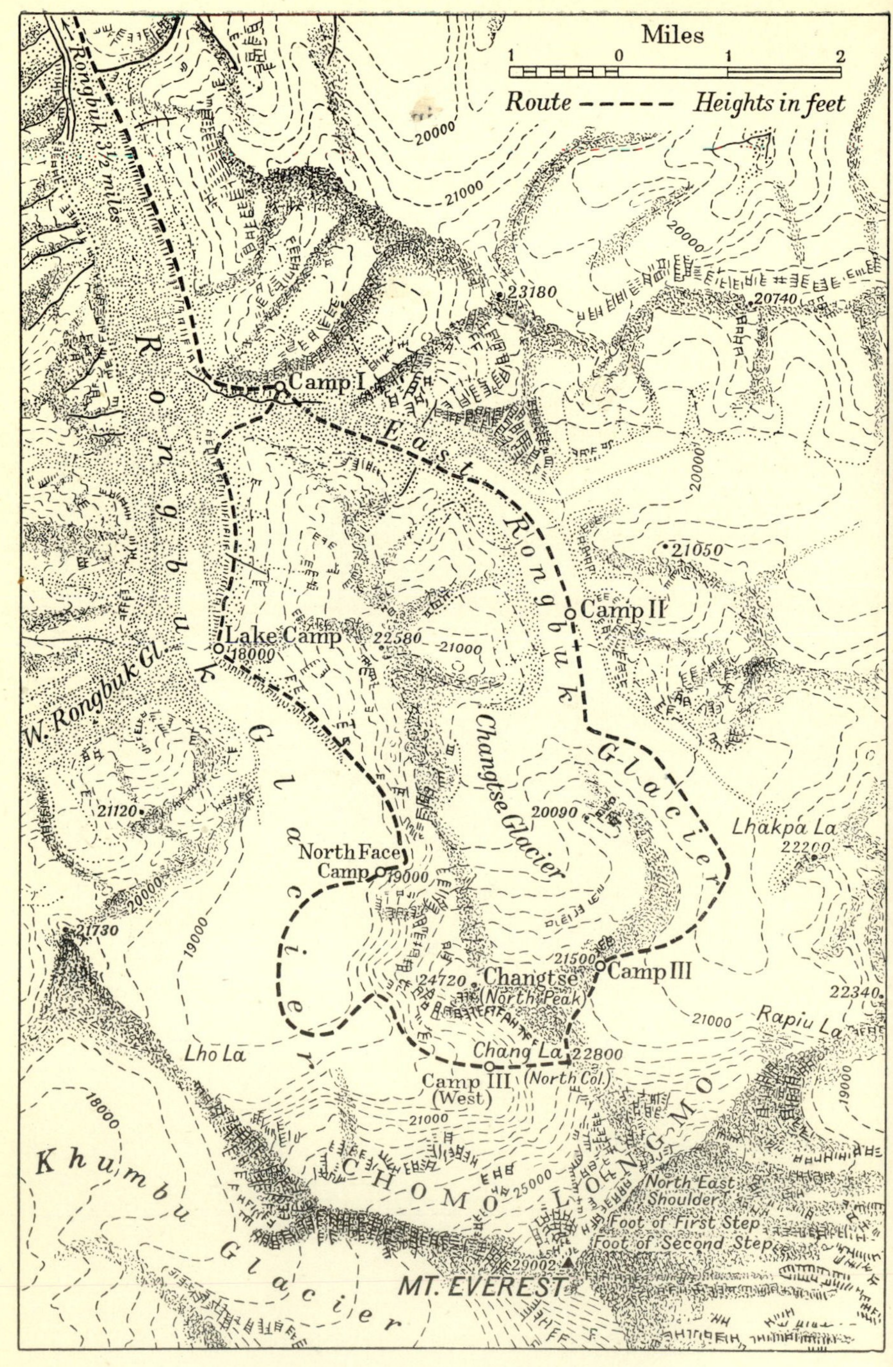

Mount Everest, the northern approaches

we slept that night, is merely a caravanserai; there is no village. Once again we changed transport so that the buildings were crowded with a double number of men and animals. After supper a hue and cry was raised by Lhakpa Tsering who had spotted a man making off with a load which he had taken from the dump in the courtyard and carried over the flat roof of the room where the Sherpas were sitting. The man escaped, but the load was recovered, and for the rest of the night the Sherpas took up their quarters on top of the baggage. According to Karma Paul this was a notorious place for thieves, but he omitted to mention it until the fact was self-evident.

Owing to the fresh transport, the third change between Tengkye Dzong and Shekar Dzong, we did not leave until 9 o'clock with our fifty-five donkeys, one pony, and two zos. Half an hour later, the wind got up and continued strongly throughout the day so that when I got in with the transport at 5.30 o'clock I was glad to join the rest of the party crouching miserably behind a stone wall which was the only shelter available. On this march we passed five very big chortens each about 30 ft. square, built up in steps like the Pyramids to a height of about 20 ft. Four were placed at the corners to form a square, and one in the middle. Karma Paul's explanation was that they are prisons for the spirits of witches. Every witch in Tibet has to pay a fee of Rs. 10 to the church so that her evil spirits may be exorcized and shut up in these chortens and the witch rendered harmless.

This Phung Chu valley was remarkable for the unusual feature of scrub growing in the river-bed and for the number of ruins and deserted villages. In these parts of Tibet it is rare to find any growth worthy of the name of brushwood or scrub except the willow groves which have been planted and cared for by man; on very favoured slopes juniper may sometimes be seen. Owing to the dry climate, walls of sun-dried brick or *pisé*, of which most are made, last for a very long time even if unprotected by a roof, and it was therefore difficult to come to any conclusion about the age of the numerous ruined buildings and chortens. On one of these we found an interesting burnt tile and an earthenware cup of a different type from those now in use locally. The commonest cause of a deserted village is the failure of the water supply on which the irrigation of the fields depends. The source is often merely an old snow-bed high up in some gully into which little sunlight penetrates, and consequently its duration is very uncertain.

On 31 March we reached Shekar Dzong where we camped in the willow grove belonging to one of the two Dzongpens. The district of Shekar is a large one divided into north and south with a Dzongpen for each, and the town, which is the headquarters of the district, is as remarkable as it is important. Village, monastery, and fort, in that order are built one above the other on a steep rock which rises for nearly a thousand feet straight from the wide, flat valley floor. 'Hill of Shining Glass' is the Tibetan name for it; a name which is by no means inappropriate when the place is viewed from a distance on a sunny day, for the rock has a peculiarly lucent appearance and all the buildings except the fort are whitewashed. The fort, perched on the very summit, is disused and is falling into decay. As there is no water there its greatest value must have been as a watch-tower, but now no one ever visits it except the monk who daily climbs to the topmost wall to burn juniper, the Tibetan's incense. We climbed up there in order to have a look at the mountain, but the day was dull and visibility poor. Descending from the fort by a new route which brought me out on the roofs of the village I was attacked by two dogs and scarcely succeeded in withdrawing unscathed behind a brisk covering fire of stones. To meet a dog in Tibet is to be attacked, but fortunately there is never any lack of stones to repel these attacks.

Besides allowing us to camp in his willow grove and giving us a hundred eggs, a couple of dried sheep, and six bags of *tsampa*, the Dzongpen invited us to spend the day with him. There is something spacious about Tibetan hospitality in keeping with the spaciousness of that vast country. You arrive at your host's house before lunch and remain there eating and drinking—mostly drinking—until after dinner without any perceptible pause. Feeble folk, like the coneys, that we were, we begged for a short break between the afternoon and evening session; so when lunch had come to a rather premature conclusion at 4 o'clock, we retired to our tents for a rest before facing the music again at 6 o'clock. One of the principal features of a Tibetan feast is the 'chang' girl, or girls, whose job it is to see that the guests drink up and that their cups are always full to the brim of 'chang' or Tibetan beer. This drink, which is pleasant and

7. Natural warm spring near Jikyop (p. 42)

8. The approach to Shekar Dzong (p. 44)

not very strong—it has been likened to slightly alcoholic barley water—is found in seemingly unlimited supply in most Tibetan households, rich or poor. It is brewed from barley which is the staple crop of Tibet, and I believe half the annual barley crop, or it may be two-thirds, is consumed as beer, the remainder being eaten as *tsampa*. Some travellers who have recorded their dislike of *tsampa* (of whom I am not one) may think it a pity that the whole crop is not turned into beer, and since the best medical opinion says that alcohol is food, no harm would be done. But at the evening session of the feast in question the drinking of this innocuous 'chang' was superseded all too quickly by that of 'arak', a spirit distilled from the 'chang', and of course bearing a very strong resemblance to hooch or bootlegger's whisky. Moreover, it was drunk from larger cups which the 'chang' girls were equally assiduous in filling. The Governor of North Carolina would have had no cause to remonstrate with his fellow Governor of the South, that profound thinker—there was no time at all between drinks. In order to make time we took advantage of a Tibetan drinking custom by which, if the guest chooses, he may challenge the 'chang' girl to sing a song which she is then obliged to do before refilling his cup. Unluckily there was a footnote to this sensible custom of which we were not aware, which is that the challenge having been accepted and the song sung, the guest is then obliged to drain the cup at one draught. So we were hoist by our own petard. The songs sung were all improvised. That in response to Odell's challenge likened him to the reincarnation of a god; Warren was told that the women of the household were not in immediate need of his services—while that to the address of Karma Paul was unprintable. It was late when we got back to our tents amongst the willows, where for some reason or other best known to themselves, the Sherpas insisted on taking off our boots for us before we went to bed.

Smythe caught us up here next day, thereby escaping this sandbag. He had with him Pasang and Rinsing whom we had left behind sick, and also Ongdi and Nurbu with whom he had travelled the previous year. This brought our strength of Darjeeling men up to sixteen, while one more Sherpa who was wandering about loose in Tibet, also attached himself to us here. On 3 April we crossed the Pang La (17,000 ft.), obtaining a remarkable panoramic view of

Everest and the neighbouring giants from the summit. There seemed to be a sprinkling of fresh snow on the North Face.

Our camp at Tashidzom in another willow grove was one of the pleasantest of the whole march. A surrounding stone wall kept the inquisitive villagers at a respectful distance; and in the absence of the usual afternoon wind, we could lie about on the grass, basking in the rays of the declining sun, watching Angtharkay cooking our supper, or soldering leaking oil drums. In fact, every prospect pleased, but man, as usual, was vile; for presently there was a hubbub in the kitchen tent and a wretched, beggarly, half-naked old man was dragged out still clutching the bag of sugar which he had been trying to steal. He was forthwith lashed to a tree while Angtharkay, thoughtfully putting some knots in a length of rope, suggested charitably that considering the subject's age and infirmity one hundred stripes might be almost sufficient to meet the case. The headman, whose servant the thief was, happened to be in the camp and flung himself at my feet begging for mercy as earnestly as if his own back were in peril. I was quite ready to let the man off, for he was already frightened out of his wits, but unfortunately I had just inflicted a heavy fine on two of the Sherpas who had stayed behind in Shekar dead drunk and had only caught us up that day; while Angtharkay, like Thwackum, was for doing justice and leaving mercy to heaven. However, we finally handed the man over to his master who promised to punish him.

On 6 April we reached Rongbuk, ten days earlier than the earliest of previous expeditions. In view of the awful warning of 1936 we had taken the precaution of being on the spot in good time; those with an over-keen sense of humour will appreciate the fact that after all this forethought, we were to be caught out by still stranger behaviour on the part of the weather.

CHAPTER V

ON THE GLACIER

And now there came both mist and snow
And it grew wondrous cold. COLERIDGE

COMPARATIVELY mild though the conditions on the march had been, wind and dust had taken their usual toll; all, except Shipton, Smythe and myself, had sore throats, coughs, or colds. At Rongbuk, Warren's sore throat had developed into influenza, while Shipton took to his bed with some stomach trouble. If the facts are investigated I think it will be found that we suffered no more from these kinds of ailments than other Everest expeditions. Coughs, colds and sore throats, accompanied sometimes by loss of voice, are a nuisance; a bad sore throat might prove to be a serious handicap on the final climb as it was to Somervell on the occasion of his great ascent with Norton to 28,000 ft. in 1924. But by the time we got on to the mountain itself at the end of May, most of us had recovered, while in some cases a loss of voice might be considered a positive gain for the party as a whole. Influenza, if that was the disease which attacked three of us, is more serious because it produces marked weakness; but as neither this nor the other complaints are preventable or even curable by the combined skill of the medical profession under the comparatively mild conditions of an English winter their occurrence on Everest expeditions is not really very startling.

Critics of the small light expedition like to attribute any illness of the members to their inadequate unscientific diet. Personally, I incline to the belief of the food reformers that much of our modern ill-health is due to faulty feeding—to the almost complete absence of fresh, natural food in the average man's diet in favour of preserved, processed, denaturalized food. But the effects of unbalanced faulty diet are long term—general ill-health is the result of months or years of wrong feeding—and it is ridiculous to blame our diet for the coughs, colds, sore throats and influenza which afflicted us at this early stage, a mere three weeks away from the plentiful food of Sikkim. Three weeks on bread and water would hardly have had this effect; indeed, it is well known, or should be well known, that coughs and colds are one result of over-eating, especially of starches and sugars.

I am not one of those who decry the mountain as unimpressive, shapeless, or even ugly—what some mountaineers rudely call a cow peak. Seen from Rongbuk it looms up magnificently, filling the head of the valley. The final pyramid, with or without its streaming banner, is a glorious thing; the face looks what it is—steep; and the two great ridges seen now in profile would make any mountaineer's heart leap. Were it 20,000 ft. lower it would still command respect and incite admiration.

When we arrived, the mountain was of course black, the snow having been blown off the upper part by the winter winds; so too, and this was unexpected, was the North Peak, 5,000 ft. lower, which in the numerous photographs that have been taken of it seldom appears as anything but a snow peak. A wind, which from the rapidity of movement of the cloud banner might be estimated at something like force 8, was still blowing up there. Down at Rongbuk there was not much wind.

Next day Kusang and Sonam Tensing came over from Sola Khombu with men and food, as arranged: forty-five Sherpas, and 1,300 lb. of rice, atta, kodo and potatoes. As 7 April was the appointed date for meeting us this was good staff work, but we had asked for only thirty men. We took them all on, however, and after a day devoted to ceremony at the monastery began moving loads up to the old Base Camp. Sola Khombu is the district in Nepal from which most of our Sherpas come, and, except for a few who have settled in Darjeeling, they return there when an expedition is over. It is not very far away from the south side of Everest, but the journey to Rongbuk over the 19,000 ft. pass, the Nangpa La, takes about a week. Later in the season even women and children cross this pass; relatives of some of our men, for instance, came over in June to see them and to pay their respects to the abbot of Rongbuk whose fame as a very holy man is widespread. The pass is therefore an easy one; but I was told that no animal is ever brought across it because the guardian spirit of the pass takes the form of a horse and any other four-footed animal which presumes to cross is instantly struck dead.

The abbot, who is now an old man, and seldom moves from his seat of audience behind a trellis-work grille, seemed pleased to see us all again. He gave us a large meal, his blessing, and some advice. An earth tremor had been felt there in February which was thought might in some way have affected the mountain; we were warned to be careful, especially Lloyd, who in spite of his superior beard was considered too young for such an adventure. We had the pleasure of watching our macaroni being made by the grimy but accomplished hands of the monastery cook before it was served to us in bowls with meat and sauces. Our bowls were replenished so many times before we finally laid our chopsticks on top of them as a sign of repletion, that I felt constrained to ask the abbot, who sat watching with pleased amusement, what the punishment was for greed in the next world. To which he made the answer that the greedy receive their punishment in this world, and very quickly. Our present to the monastery was a very fine altar cloth, and the Sherpas gave rupees (provided by us) which were offered stuck upright in a bowl of uncooked rice with a piece of sugar on top.

On 9 April we made our first carry, taking 57 loads to the old Base Camp where we left two men. There was much ice about. On the way back it snowed and blew hard—conditions which, attending our first move in the game, were regarded by the Sherpas as of evil omen. They betook themselves to a very holy man who lives alone in the severest simplicity in a small house close to the camp four miles above Rongbuk. His serene and cheerful countenance filled me with wonder. Each man listened with reverently bowed head to the few words said to him and received with manifest joy some large, repulsive-looking brown pills. The march from Rongbuk to the old Base Camp takes but an hour and a half, but it was with the greatest difficulty that the men were persuaded, against all precedent, to make two carries on the next day. This completed our first move, and by 5 o'clock on the 10th we were all installed there except Warren whom we had left with Karma Paul in a house adjoining the monastery. The previous night there had been a violent storm with zero temperature resulting in the North Peak turning white. Everest itself remained black, either because less snow fell up there or because the wind blew harder.

Our porter corps was now twice as large as the thirty we had anticipated. We had neither tents nor clothing for so many and since we were in no hurry now to reach the mountain—for owing to the prevailing cold it seemed unlikely that anything could be done until May—we decided to dismiss some. So after making two carries to Camp I, we picked out fifteen of the Sola Khombu men and sent thirty home, thus leaving us with a total Sherpa strength of thirty-one. Sorting out the sheep from the goats was largely guesswork; only two or three were old hands whom we knew, and for the rest we had only appearances to go by which in the case of Sherpas is not always a reliable guide; nor was the advice offered by their friends amongst our Darjeeling men entirely disinterested. All were pathetically eager to stay. The thirty who were dismissed with a gratuity consoled themselves further by taking with them five of our thirty precious pairs of porters' boots—a serious and unexpected loss, for with only Sherpas in the camp I had imagined that precautions against theft might be relaxed. By borrowing spare boots from the Europeans some of the shortage was made good, but three of our thirty-one porters never had anything except their Tibetan boots and consequently were unable to go any higher than Camp III.

Partly as a result of this boot shortage a regrettable incident has to be recorded. As Angkarma, who was Odell's servant, had been speechless with laryngitis for the previous fortnight and did not look robust, it was thought kinder to send him home and retain someone who was more likely to go high. With this idea in mind it became essential to have the boots which had been issued to Angkarma at Kalimpong, which were now of far more importance than he was. But with these the owner very resolutely declined to part. A scene ensued in which Angkarma lay on the ground, making what noise a person in his condition could, but managing to name in a hoarse whisper such a stiff price for the boots that he easily won the day and we got neither the boots nor a better porter.

The next day, 13 April, we occupied Camp I. We took with us all we should require for a month's stay on the mountain, and maintained neither lines of communication nor any base except Rongbuk where Karma Paul remained with the rest of our stores. I spent a long afternoon issuing kit to the fifteen new men, and

Primus stoves, oil, and cigarettes to all. From now on cooking for the whole party had to be done on stoves.

Someone had to go without boots, and since Sonam Tensing was mainly responsible for recruiting the Sola Khombu men, amongst whom there were apparently some thieves, I made him one of the scapegoats. In any case it is usually a waste of money giving kit to the 'Foreign Sportsman' (the name he went by among us) because he has a habit of wearing his own clothes throughout the campaign and retaining ours for subsequent sale or barter. I remember in 1937, when we last travelled together, he favoured a natty double-breasted grey summer suiting with very nice thick white stockings worn outside the trousers, while in 1938 I never saw him in anything but the clothes he came in from Sola Khombu which kept him just as warm as our elaborate sweaters and wind-proofs. His attitude to clothes was thus very different from when he began his mountaineering career in 1935 and first earned the sobriquet of the 'Foreign Sportsman'. He joined us then at Camp II on this same glacier, having blown in unasked and unheralded from heaven knows where. We admired his enterprise, so we took him on and gave him a complete rig-out which he never discarded until Darjeeling was reached some months later. In July or August, miles from any snow, he would be seen equipped as the complete mountaineer in wind-proofs, gloves, puttees, snow-glasses, Balaclava helmet, as though about to face some terrific blizzard. He is a good and likeable porter with many attractive features, but amongst them I would not put the deep bass voice with which he croons endless Buddhist chants. The patience shown by his comrades in bearing with not only these, but also with his faculty for evading camp chores and a not unwise regard for his own comfort, led us to suppose him an unfrocked monk.

Oliver and I accompanied the first relay of loads to Camp II which was reached in four hours on a fine, cloudless, almost windless day; that is to say windless so far as we were concerned, but the banner streaming from the top of the mountain showed only too clearly what was happening there. Now that we had thirty-one men, of whom—owing to camp duties—only twenty-nine were usually available for carrying, four relays were necessary for every move of camp.

There seemed to be a disgusting amount of stuff to shift. No one could have accused us of lightness or mobility. The slogan 'No damned science', if raised at all, had evidently not been heard by Odell. We had amongst our scientific equipment our old friend the glacier drill and a machine which, in return for the slight trouble of winding up, recorded the relative humidity. The results were so unexpectedly various that one concluded the thing was only guessing. Then we had batteries of thermometers, and they were all needed; for in order to register impressively low temperatures they have to be whirled joyfully at the end of a piece of string, and naturally a lot of them flew away. Our library, too, was a weighty affair. Shipton had the longest novel that had been published in recent years, Warren a 2,000-page work on Physiology. Odell may or may not have had a book on Geology, but he himself daily wrote the equivalent in what he humorously called his field notes. Oliver, no doubt, had Clausewitz on the *Art of War* and Lloyd a text-book on Inorganic Chemistry, but I had carelessly omitted to bring either a standard work on Mountaineering, Mrs Beeton, or the *Chemistry of Food*. Moreover, I had to use all my persuasive art to induce Angtharkay to leave at Rongbuk a great bread-making oven, though he well knew we should seldom have any fire for baking and though he had often watched me turn out perfectly good bread from an ordinary 'degchi' or cooking pot. One of the troubles of having a real cook is that he always wants a great quantity of impedimenta. If you curtail it and then curse him for any shortcomings he can always retort that it is your own fault for having thrown out some particular implement or ingredient. But there was never any trouble in this respect from Angtharkay, for whom cooking was a diverting side-line, and who could manage on very little and manage uncommonly well. So long as camp cooks have not got the five-course-dinner complex, from which all Indian cooks suffer, and content themselves with one simple dish such as they themselves would eat, then one can be sure of having a square meal.

On Good Friday we gave the men an off-day with the result that their stoves, filled in the morning, were empty before evening. Owing to leakage and prodigal usage, oil was a constant source of worry; but having managed to buy eight gallons, left there in 1936, from the monastery, we finally finished up with several gallons in

9. The Gompa of Shekar Dzong (p. 44)

10*a*. Transport negotiations at Shekar Dzong
(p. 45)

10*b*. Outer courtyard of Rongbuk Gompa
(p. 48)

hand. The rest of us lay about, played chess, or read the less technical portion of our curiously assorted library. This included *Gone with the Wind* (Shipton), *Seventeenth Century Verse* (Oliver), *Montaigne's Essays* (Warren), *Don Quixote* (self), *Adam Bede* (Lloyd), *Martin Chuzzlewit* (Smythe), *Stones of Venice* (Odell), and a few others. Warren, who rejoined us that day, besides his weighty tome on Physiology—in which there were several funny anecdotes if one took the trouble to look—had with him a yet weightier volume on the singularly appropriate subject of Tropical Diseases. Perhaps he wished to be prepared for all possibilities, or, since the temperatures were around zero, he found it a vicarious source of heat and thus supplied the answer to Bolingbroke's question:

> O, who can hold a fire in his hand,
> By thinking on the frosty Caucasus?
> Or wallow naked in December snow,
> By thinking on fantastic summer's heat?

Camp II was occupied on 18 April. It was in the glacier trough on the 1936 site, but of that camp there was scarcely a trace. No one walking up the East Rongbuk glacier would dream that there in the last twenty years more tins had been opened and more rubbish dumped than in any comparable area in India. It is, indeed, fortunate for us that the Tibetans are such thorough scavengers; the camps on the glacier would be even grimmer than they are if the accumulated debris of all the past expeditions still lay there. Returning through Sikkim this year I happened to visit the base camp on the Zemu glacier used by the Germans for their *two* attempts on Kangchenjunga; the amount of rubbish strewn about was reminiscent of the day after a Bank Holiday at one of England's more popular beauty spots.

Immediately on arrival at Camp II I went to bed with an attack of influenza following hard upon the preliminary warning of a sore throat. One carry was made to the old Camp III situated just on the corner where the glacier bends round towards the North Col, and then we sent all but three of the men down to Rongbuk where they could have a day's rest before returning with some more loads. The thermometer was now recording temperatures of 46° and 47° of frost at night. On the 22nd I felt strong enough to go down to Rongbuk, but before I went we held a discussion about plans; or rather the others discussed while I listened, for since going sick I had completely lost my voice. Whether the others noticed any departure from the normal I cannot say. The decision reached was that after having a look at the North Col slopes, Shipton and Smythe were to retire to the Kharta valley while the rest of the party were to go up to the Col, and, if conditions warranted, proceed with the establishing of camps on the mountain. Both Shipton and Smythe, who had had most experience, were emphatic that the best time to attempt the mountain was the end of May or the beginning of June —that is just before the onset of a normal monsoon if there is such a thing in these parts. They were strongly averse to making an earlier attempt. As they were considered our most likely pair it was decided to reserve them for this main attempt. Meantime, we others would have a look at the North Col and decide whether conditions allowed of commencing operations on the mountain. Shipton and Smythe were not to wait for our report but were to proceed direct to Kharta from Camp III and there fatten up for the kill.

Accompanied by the 'Foreign Sportsman' I went down to Rongbuk in one march, passing on the way the porters who had that day begun their return journey. I put up with Karma Paul in his quarters by the monastery gate, and after a strong dose of hot 'arak' and butter at bedtime felt very much better by next day, which we passed playing piquet with satisfactory financial results to myself. I was told that the porters had also dosed themselves liberally with 'arak' when they were down, but not medicinally. Karma Paul's servant, Pensho, cooked for us. He was quite black with smoke from his cooking operations and had completely lost his voice although he had never been above Rongbuk. This versatile lad grooms and rides racing ponies in Darjeeling in ordinary life— sometimes for Karma Paul, who himself is a bit of a racing man. Pensho, by the way, has his name tattooed on his forearm very large and legibly.

On the 25th we went back to Camp II in one jump, the 'Foreign Sportsman' doing a man's job by carrying up about 80 lb. Twenty porters were there, six of them sick, but none wanted to go down, so next day we all went up to Camp III where the rest of the party now were. On leaving the moraine trough for the open glacier we were upon bare, slippery ice; but powder snow, collected thinly in cracks and hollows, allowed us to progress without skidding about or having to nick steps. Only Pasang was in Camp III; the others could be seen half-way up to the North Col whence they presently returned. They reported that there was much ice on the slopes, which had obliged them to cut steps as far as they had gone, but that the route would probably 'go' all right to the top; they also reported that it was extremely cold—a fact which I had already discovered for myself without leaving camp. The whole party were now, 26 April, at Camp III with a month's supplies, and the question was what to do next.

In view of the prevailing wind and cold the general opinion was against proceeding with the establishment of the higher camps at present. None of the party, except Shipton, was at this time free from some minor ailment. Oliver's cold had become chronic, Odell had a cough, Smythe a sore throat, Lloyd and Warren colds, while I was merely convalescent. For these ailments drugs are of no more use on Everest than they are anywhere else; the only remedy is to go down. In accordance with the original plan, Shipton, Smythe, Oliver, and nine porters, started for the Kharta valley on the 27th. The rest of us intended waiting for a bit, partly to see whether the weather would become any warmer, but principally in the hope of receiving a mail which we thought might arrive on 1 May. But extremely low temperatures, combined with inactivity, were having no good effect on us either physically or mentally, so on 29 April we too set out for Kharta with thirteen porters.

On account of its low altitude (11,000 ft.) the Kharta valley was the place where we should most quickly throw off our manifold infirmities. The only alternatives to this move were to stay and carry on with the attack or to go down to Rongbuk, or Camp I, which would have been nearer to the mountain but less beneficial to health—Rongbuk is dusty, dirty, and 16,500 ft. above the sea. Before we left England, General Norton had advisedly warned me against the danger of committing the party to an early attempt with the possible result of putting most of us out of action with frost-bite. In 1924 his own party had suffered severely from cold weather early in May even at Camp III; twice they had to retreat down the glacier with the result that when favourable conditions at last arrived the party were worn out with the battering they had received. So severe were conditions this year at Camp III at the end of April that frost-bite was not so much a possibility as a certainty for any one on or above the North Col. Had we decided upon going to Rongbuk in order to be nearer the mountain nothing would have been gained. The cold and the wind continued unabated for the next six days and then on 5 May snow fell heavily and continued to fall daily for the next week. After this the mountain was never in climbable condition. That lull, on which all depends, between the dropping of the north-west wind with accompanying milder weather and the first heavy monsoon snowfall never occurred. I think it is true to say that at no time this year was the climbing of the mountain ever within the bounds of possibility.

In a letter expressing his best wishes for our success a candid friend warned me that whatever befell I was not to put too much blame on the weather. Such a caution might be more widely applied. How ready many of us are to find in the weather a handy and uncomplaining scapegoat for our less successful enterprises, our ill-health, or low spirits. Dr Johnson, that arch-enemy of cant, had a hearty contempt for the man who permitted the weather to affect his spirits or his work. 'The author who thinks himself weather-bound', he declared, 'will find, with a little help from hellebore, that he is only idle or exhausted.' A weather-bound mountaineer is more common and more likely to prove a genuine case for sympathy than a weather-bound author, but if we search our hearts we may have to confess to having on occasions made the weather a pretext when in fact we were idle or exhausted, or to cover up our irresolution, unfitness, or lack of judgement or skill. Nevertheless, I think it is generally admitted that the weather factor and the condition of the mountain is of greater consequence on Everest than on other mountains. At least all mountaineers now recognize the overwhelming importance of it; but I sometimes wonder whether the layman may not think that just as a bold sailor takes the weather as it comes, refusing to be cowed by it and only running for shelter in the last extremity, so a sufficient show of resolution on our part might overcome our difficulties. But the mountaineer, obedient to his code and aware of his limitations, must be allowed to judge what is possible and what is not, what risks he can justifiably take, nor must he rely very much on luck to retrieve his mistakes. Where the prize is so great he is not likely to err on the side of caution, but we should not forget that mountaineering, even on Everest, is not war but a form of amusement whose saner devotees are not willing to be killed rather than accept defeat.

The morning of the 29th, when we abandoned Camp III, was very cold indeed, as I quickly found when standing about making up loads for the journey to Kharta. We expected to be away about a fortnight, returning to Camp III by the middle of May. We left behind nine Sherpas, who, after spending a week at Rongbuk, were to return and carry the camp to the new Camp III site about 500 yards higher up the glacier. Looking up at the mountain, seemingly in perfect condition for climbing, it was impossible not to feel some misgivings at turning our backs on it and marching away. But even here, at 21,500 ft., with no wind blowing to speak of, the cold was sufficiently intimidating to banish all regrets. The fact that we never again saw the mountain black inclines one to curse one's pusillanimity for missed chances; but it is only necessary to recall that a month later, when conditions had changed so much that the heat at Camp III was positively enervating, at 27,000 ft. the cold was barely tolerable. There is no question that at this time no man could have climbed on the mountain and lived.

CHAPTER VI

RETREAT AND ADVANCE

I'll go, said I, to the woods. COPPARD

THE route by which we retreated from the East Rongbuk glacier to the Kharta valley was that used by the party which first set foot on the mountain. In the early stages of their exploration of the Rongbuk glacier in July 1921 Mallory and Bullock realized that the two great ridges descending from Everest to the north-west and south-east were not practicable routes to the summit; apart from their great length, and whatever difficulties there might be higher up, neither of them was easily reached. There remained the north ridge, which unlike the other two did offer some hope of approach by means of the comparatively low col joining the north ridge to the North Peak. The western side of this col, which came to be known as the North Col, was examined but discarded as being unsuitable for laden porters. The eastern side remained hidden, and although an exploration of the East Rongbuk glacier was to be undertaken, Mallory and Bullock did not suspect that it would lead to the North Col; they thought the glacier draining from that side of the col must flow away to the east and south; for the quantity of water in the stream which issued from the snout of the East Rongbuk glacier hardly seemed great enough if that glacier had its source at the foot of Mount Everest itself. Their exploration of the East Rongbuk glacier which would have cleared up the topographical puzzle, but at the same time might have deprived the party of much of the interesting work accomplished later, was deferred. The time allotted to it had to be devoted to the retaking of a number of important survey photographs which previously had proved failures, and the opportunity never recurred.[1]

The party abandoned the Rongbuk glacier and moved round to the Kharta district to examine the eastern approaches to the

[1] Col. Oliver Wheeler, however, who was working on a photographic survey detached from the others, went far enough up the East Rongbuk glacier to hazard the guess that its source was beneath the slopes of the North Col, but this hint did not reach Mallory and Bullock until they had transferred their attentions to the Kharta valley side.

mountain. One glacier, the Kangshung, brought them under Makalu and the south side of the great north-east ridge; but at length Mallory, Bullock and Morshead, climbing a col at the head of the Kharta glacier, looked across a wide glacier floor leading directly to the North Col. They were on the Lhakpa La (22,500 ft.), and the glacier below them was, of course, the upper extremity of the East Rongbuk. Running eastwards from the North Peak was a ridge which hitherto they had imagined formed the head wall of that glacier, but now it was seen to be merely a buttress round which the glacier swept in an almost right-angled bend. In September, from their base in the Kharta valley, they attacked the mountain from a camp on the Lhakpa La itself. In view of the condition of the party at the end of a long and strenuous season, the inexperience of the porters, the state of the snow, and the weather, the fact that they reached even the North Col witnesses to their strength and determination.

During this year's attempt we discussed the future possibility of using the Kharta valley as a base. Its principal advantages are: a shortening of the approach march by about four days by cutting out Shekar Dzong; very much pleasanter living conditions at the base compared with those at Rongbuk; more easily obtainable food and fuel; and a local porter supply. There is no question that Rongbuk is not a pleasant place in April, May, or even any other time. It is cold, dusty, and as dirty as might be expected from the presence in a confined space of two or three hundred permanent residents with no notions at all of sanitation. Our tents are pitched on the only patch of grass; for the rest it is stones and rocks with never a tree and scarcely a flower. All food and firewood have to be brought from the nearest village a day's march away, and there is no labour available locally. Yet it is my belief that these disadvantages, which are not slight, are offset by the shortness and ease of the approach to the mountain. Having to cross an obstacle such as the 22,500 ft. Lhakpa La on every journey to or from Camp III would be a serious consideration; the Kharta side of the pass is steep, while there are many crevasses on the upper part of the Kharta glacier. Therefore, men could not travel alone; parties would have to be roped; bad weather or heavy snowfalls, which on the old route can be almost ignored, would on this route be regarded with concern; and the difficulty of getting down sick or injured men would be greatly increased. Two other points for consideration are, first whether or no the Tibetans would raise any objection, and secondly the fact that the North Face of the mountain is not visible from the Kharta side. No doubt special permission to alter our route and to base ourselves on Kharta instead of Rongbuk would have to be sought, for when the Tibetan authorities grant a passport they lay down a route and expect us to follow it. What effect if any arises from a party's having the mountain ever before its eyes I cannot say. Like most things in life one probably gets used to it, and it would be unwise to suggest that there is any effect, lest in the interests of science some future party finds itself saddled with a psychologist to study the question. But it is valuable in other ways to have the mountain always in view. I regretted later that Karma Paul was not given the job of taking a photo of the mountain at the same hour every day of our stay; such a record of the changes in snow conditions on the face would be interesting and informative, as would be meteorological records which could also be kept by whoever was left at Rongbuk. No active member of the party can take regular four-hourly observations.

Odell, Warren, Lloyd, myself and thirteen porters left at 10.30 o'clock on 29 April. Heading straight across the glacier on slippery grey-blue ice we had on our right the foot of the north-east ridge where it runs down to the Rapiu La. Odell and I were behind so that as we approached the eastern bank of the glacier we could see the others beginning the climb to the pass moving in single file up the crest of a rock spur of which the flank was sheeted in ice. Suddenly a figure, which I recognized as Warren, slid down the ice. We hurried forwards as fast as we could on the slippery surface, but before we reached him he had come to an abrupt halt among some rocks after sliding about 100 ft. I was relieved to see him sit up. No serious damage had been done but he was severely bruised and shaken. He had great difficulty in going on, but accompanied by Odell managed to do so slowly. What had happened was, that when following the rocks lying just above the ice-slope, he had stepped on a stone which was not frozen to the underlying ice and which consequently slid away.

11*a*. The Holy Lama of Rongbuk Gompa (p. 49)

11*b*. The Base Camp in Rongbuk Valley (p. 49)

12*a*. The party. Left to right: Chas. Warren, Peter Lloyd, H. W. Tilman, P. R. Oliver, F. S. Smythe, N. E. Odell, E. E. Shipton (pp. 15 and 50)

12*b*. Angtharkay and a young brother (pp. 26 and 50)

We struggled upwards in a very cold wind. I seemed to be going very badly, but took comfort from the thought that no one else seemed much better. At 2 o'clock we reached the top where we were relieved to see the large steps cut by the other party down the eastern side still intact. The slope was ice but the porters tackled it confidently in spite of their heavy loads. It was too cold to linger, but as I turned my back on the mountain to begin the descent the chief impression I took with me was the unpleasant appearance of the north-east *arête* and the amount of cloud away over Sikkim to the south-east. The glacier was not slippery ice like the East Rongbuk but afforded good going on hard rough snow. We walked rapidly down it, admiring on our right the beautifully fluted snow ridge between us and the head of the Kangshung glacier, beyond which, over a low saddle, we watched the clouds forming and dissolving upon the black precipices of Makalu. There was but little wind this side, and the fact that the glacier surface was snow and not bare ice suggests there never is enough wind to sweep the snow off.

We continued the march next morning down the left-hand moraine. After going for two hours we reached the snout of the glacier and from there followed the excellent hard highway provided by the frozen river. Warren, stiff and sore from his fall, had much difficulty in walking, so for the last mile or so he was carried perched on the broad shoulders of Nukku. We were now entering a more genial climate; grass and plants, only recently freed from their snow-covering, were beginning to show green promise of fresh growth. On 1 May we camped close to the first village, having already passed the yaks moving up to higher pastures. A recurrence of my influenza began here; but, expecting to come across Shipton's party, we pushed on down the valley to camp at length in a delightful grass meadow by the side of a big 'chorten'. Juniper wood abounded. The attractiveness of the place and my sickness made me disinclined to move next day, so here we stayed while Lloyd and Warren went on down the valley, crossed the Kharta Chu to the south side, and found the camp of the other party tucked away in a wood overlooking the Arun gorge. We joined them next day, 4 May, when I had to take to my bed with a bad cough, a worse cold, stiff legs, and a temperature.

We moved the tents out of the wood into a broad meadow sheltered on two sides by forests of birch, juniper and rhododendron. On one side of our tents a little tarn reflected the sky while on the other side, barely fifty yards away, the ground fell sheer for nearly a thousand feet to the bed of the Arun river, the roar of whose waters came faintly to our ears. Yaks grazed in the meadow, the rhododendron was in bud, birds sang, the pale green shoots of the birch were beginning to show; only an Eskimo could have regretted the ice and snow of Camp III. There was no village near us, but the Kharta Chu down which we had come is a fertile valley with many prosperous looking villages each one with its monastery, so we sent out daily foraging parties who brought back sheep, flour and eggs, although at that time of year food was scarce and consequently dear. Moreover, since leaving Rongbuk the rate of exchange had moved against us in the mysterious way exchange rates have, and now we received only 25 Tibetan 'tankas' instead of 30 for our rupees. This movement was more unaccountable than those which take place on the international market; for there, however the rate may vary from time to time, it does not vary from place to place—a pound is worth the same paltry number of dollars in Boston as in New York. To the ordinary man the only fixed point in the shifting phantasy of exchange rates is that they are usually unfavourable to him.

The Arun gorge, at the head of which we were camped, is only one of several great gorges carved out by rivers in the Himalayan ranges. It is, however, a peculiarly striking example when seen by the traveller from the Tibetan side; for he is already familiar with the sluggish meanderings of the river, apparently so easily diverted from its course by slight barriers, and has stared amazed at the seemingly impenetrable front of the gigantic barrier which it at last pierces. It is another instance of a phenomenon frequently observed: namely the proximity of deep gorges to high peaks, for the rivers often cross the range near its highest points. The Arun gorge is within 10 miles of Makalu and 20 miles of Everest; the gorge of the river Gori, in Garhwal, within 12 miles of Nanda Devi. Other instances are Namcha Barwa and the Tsangpo gorge; and Nanga Parbat whose foot is almost washed by the Indus. The principal branch of the Arun in Tibet is the Phung Chu, which drains the whole area stretching westwards for 200 miles from the point where our route enters Tibet, and northwards to the watershed of the Tsangpo or Brahmaputra—from the Sebu La, near Tangu, to Gosainthan, a 26,000-ft. peak nearly 80 miles west of Everest. It forms the principal tributary of the Kosi river of Nepal which in turn flows into the Ganges. How these great gorges came into being is a fruitful source of geological speculation and controversy. Several theories have been evolved to account for the phenomenon of a river flowing directly across a great mountain chain, the two most favoured being either the 'cutting-back' by the river, or the antecedence of the river to the mountains, with the maintenance of its original course as the mountains were raised. The parallel question of the antecedence of hens and eggs inevitably suggests itself to the irreverent.

Though our meadow was sheltered by steep wooded slopes and the tents themselves pitched in the lee of a clump of trees, yet it was one of the windiest of spots; a constant blast swept up the gorge like steam from a safety valve. On our third day there the wind brought with it clouds and rain, which fell as snow on the heights overlooking the gorge. Clouds also formed away to the north, but overhead was a patch of blue sky where two opposite wind currents battled for mastery. It was at this time, had we known it, that snow was falling heavily on the mountain, but we flattered ourselves that the unsettled weather we were experiencing was only what might be expected when the moisture-laden airs of the plains, drawn up through the funnel of the gorge, met the cold winds of Tibet.

Nobody was very active during the six days spent here. The more energetic went for short walks, the most favoured being to a point of vantage on the very lip of the gorge where you could sit on a rock with your feet dangling over the water a thousand feet below. Those with a mathematical turn of mind exercised it by hurling stones into the river and estimating the height from the time they took to reach the water. A track carved out of the cliff on the opposite side aroused such interest that we made plans for going down the gorge and home through Nepal, knowing very well that we never should. I remained for the whole time on my back except for tottering a few yards on the last day to see that my legs were still functioning—a sort of training walk preparatory to the march back. So did Smythe, but for a different reason; for his belief is that a rest should be a rest and that only essentially unavoidable movements such as lifting a cup or reaching out a hand for food should be undertaken. A pleasant custom was started of dining together at night round a huge fire lit in the midst of the clump of trees. Interminable arguments on serious subjects were carried on, but sometimes snatches of livelier conversation were wafted to me on the breeze as I lay in my tent. Thus:

'...but since there is an entire absence of lacustrine deposits....'

'...good, but do you know the one about the young man of Baroda?'

'...I thought the Turner pictures frightful.'

'...brought Peacock instead of Montaigne. Old Montaigne is delightfully bawdy but Peacock's cleverer and his descriptions of food are grand.'

'Food, my God! Mutton every day since Gangtok....'

'...and suddenly a tall Frenchman with a green face and a thermometer suspended round his neck was sick on the floor of the hut.'

'...have you heard the story of Simpson the strong guy?...'

'...lucky to get all that stuff at Rongbuk. Bath Olivers, Pumpernickel, pickled onions....'

'...probably metamorphosed rock with....'

'...never bothers about lunch—thinks it sissy....'

'...The Rembrandts gave me the willies....'

'...saw me putting a hunk of bread and cheese in my pocket for lunch and offered me a tiffin coolie to carry it....'

'...Did the Meije and got back to Grenoble by....'

'...ah well, I said we should starve. Don't eat all the jam, its all we've got....'

'...by taking levels to the river terraces....'

'...four different kinds of marmalade in '36....'

'...write my field-notes up by the light of half an inch of candle. Good night.'

On 10 May five of us left to return by the Doya La to Rongbuk; this was an easier way than the Lhakpa La and I hoped to get a pony to ride. We left behind Shipton and Smythe who were to return by the Lhakpa La and meet us at Camp III on the 20th. Shipton had a sore throat which proved to be the beginning of a mild attack of influenza. Our first march was a short one of about five miles to a village at the foot of the Chongphu valley which leads to the Doya La. I felt weak and was bothered by spasms of coughing, but unfortunately I was unable to hire a pony; we were told that everything on four legs had gone away to take part in some mysterious 'races'. However, we bought sixty eggs and a sheep which were more useful. The Sherpas have a curious dislike of slaughtering a sheep. They love meat, especially the messier bits of the stomach; they enjoy slicing off chickens' heads, cutting up a dead sheep, cleaning the intestines, removing the brains, but they jib at the actual killing of the animal. If no one else will oblige they cast lots amongst themselves, while the loser is usually so upset that he bungles the job and makes two or three half-hearted blows with the kukri instead of one, before severing the beast's head. No doubt if we had to slaughter our own meat there would be many more vegetarians than there are.

Our valley was very pleasant; the stream which bubbled down it was so clear, and brown, and merry, that it reminded me of some Lakeland beck. No great glacier feeds it; its source is a lake which in turn draws its water, not directly, from a small glacier. I think no glacier-fed stream can appear friendly as other streams do. It is neither good to drink, pleasant to bathe in, nor easy to cross; over it hangs the dank smell of a grave dug in clay, and it has the cold, grim, forbidding air of its parent. We were too early to see many flowers, but later on this valley is full of roses, clematis, and primulas; for on this side of the Doya La the climate is very much damper than on the north. Only the children remained in the villages to take any interest in our passage; all men and women were out in the fields busy sowing barley and guiding the water from the furrow on to the sown plots.

We crossed the Doya La (17,000 ft.) on 12 May, a dull day with a threat of snow in the air. Odell and I, who mistook several likely looking depressions for the real pass, steered a rather erratic course, but we caught up Warren and the porters below the true pass and spent two hours and a half very pleasantly, brewing tea. To be more accurate, brewing one small cup of tea; for so exiguous was the fuel supply that it took all that time to boil the water. We were so long about it that Lloyd and Oliver, who had also visited a false pass, came back to look for us thinking we had made camp. There was snow on the way up and more on the north side of the pass which we crossed in a snowstorm. On the top Odell took advantage of the excellent light afforded by the mist and falling snow to take a round of photos of topographical, or perhaps, meteorological interest. An hour's swift run down brought us to a wood where we camped.

And so next day to Chodzong where we joined the road from Tashidzom which we had traversed in April on the way up. Here we heard news that Purba Tensing had just gone through with the mail and were put into a fever by the thought that Karma Paul, ignorant of our whereabouts, might send it straight up the glacier. Hard though one might strive to persuade oneself that it was just as cold now as in April, there was no denying that the weather was positively mild. The glimpse of the mountain which we caught as we moved off next morning on the way to Rongbuk confirmed our worst fears. It was perfectly white, and from the summit a plume of cloud trailed lazily from the south-east to the north-west carrying to us an unmistakable message. I did my best to convince myself and the others that it was not Everest that we were looking at but some other mountain. Those amongst us who habitually took a gloomy view were certain the monsoon had begun, but we only voiced this fear for the pleasure of hearing it scoffed at by those with more robust minds. It was still early in the season and the hope that the north-west wind would re-establish itself and clear the mountain was not unreasonably entertained by all. It seemed a reasonable hope then; but it was no light dusting of snow at which we were looking but a heavy coating which grew daily thicker.

At Rongbuk we heard from Karma Paul what had happened. How on 5 May snow had fallen and continued to fall daily for the next week. In his diary for 5 May he writes: 'The mountain is as white as anything....' We found our mail, devoted a day to reading it, and on the 16th started once more for the mountain.

CHAPTER VII

ADVANCE AND RETREAT

Hoping for the future, tormented by the present.

At Camp I there was a lively exchange of news when we met Ongdi and his men whom we had left behind. They had just returned from up the glacier where they had been busy shifting Camp III to the higher site. They reported a foot or more of snow on the glacier. It was calm and mild as we moved up this time, and when we reached Camp III on 18 May we realized how drastically conditions had changed; a foot of snow covered both the ice of the glacier and the rocks of the moraine, while water was found lying about in pools ready to hand whereas before we had been obliged to melt ice. For some reason or other—we preferred to put it down to the heat—we all found the 500 yards going up the moraine to the new camp site particularly trying. I checked the stores and found we had now accumulated sufficient food for five weeks.

Very early next morning clouds began pouring up from the south over the Rapiu La bringing in their train snow which fell intermittently throughout the morning. This was not encouraging, but Odell and Oliver went a short way up the North Col slopes to find the snow in good condition. Lloyd and Warren spent the day trying unsuccessfully to mend one of the 'closed' type oxygen apparatus which had gone wrong, while I offered advice from the depths of my sleeping-bag where I lay with limbs which ached as though they had been racked. This, I thought, heralded another attack of influenza, but nothing came of it. Lloyd was not feeling well and in his case influenza developed the next day, 20 May, so that he was unable to accompany us when four of us and four Sherpas began the task of making the route up to the North Col. The temperature that night was 1° F.

As at first we had intended taking laden porters with us, Oliver went off early with two men to kick steps, leaving us to follow an hour later escorting the porters. After prolonged discussion we decided that it would be wiser to leave them until the route was nearer completion, so Odell, Warren, myself and two Sherpas, started rather late. The route we chose was a fairly central one in preference to a slightly easier line more to the right which, however, involved a very long traverse across the place where Shipton and Wyn Harris had nearly been avalanched in 1936. Half-way up there was a rather critical place where we had to cross the line of fire of a threatening ice cliff from which a mass of big blocks had already fallen; and for a time we were in a sort of funnel which was rather too well adapted to act as an avalanche chute had one been started from above. Actually, we never had any trouble at all here, but on this first morning, right at the foot of the slope, we late-comers were reminded of the advantages of an earlier start by the fall of several large blocks of ice just to our left.

The snow was good and the climbing so easy that Oliver did not call upon those behind to take any share in the work until within about 300 ft. of the top. There the slope steepened abruptly and broke into ice cliffs; in consequence we were forced to traverse to the left for about a hundred yards before the slope eased and allowed the route to be finished by a direct climb to the top. The snow on the traverse was steep, loose, and soft. A rope would obviously have to be fixed here for the laden porters, so Oliver wisely waited for the rest of us to come up before embarking on it with his two Sherpas, who were carrying one end of a 300 ft. line for fixing, and some long wooden snow pitons like cricket stumps. They had not proceeded far before they became bunched together with the result that the snow at once avalanched. The leading man, who was a Sherpa, was beyond the cleavage, which was about two feet deep, and the light line which we were paying out to him got mixed up in their own climbing rope so that we easily held them before they had been carried far down. They climbed back to where we were sitting, on a snow boss, which made a convenient and safe stance at the beginning of the traverse. Then, having thus 'tried it on the dog', Odell and I, secured and doubly secured by many ropes, took over the task of cutting and stamping out a continuous track in the steep snow. The first part of the route at least had now been made safe.

Oliver's party were sufficiently safeguarded to prevent any serious consequences arising out of this avalanche which at the time none of us regarded very seriously. I did not report it, but there was

13*a*. Eastern descent from Lhakpa La; Peter Oliver and Sherpas; Everest in cloud (p. 61)

13*b*. Amid the pinnacles of the East Rongbuk glacier (p. 53)

14. Everest from the Lhakpa La, 22,200 ft. (pp. 59 and 61)

a leakage of news somewhere and later I was disgusted to find the popular Press had got wind of it and had related the story with their usual happy accuracy. Glaciologists will be interested to know that the party had been 'caught by the tail of the glacier', and meteorologists that they were 'nearly carried away by the monsoon', both these nasty mishaps taking place on the 'North Column'. But it would be ungenerous to withhold praise where praise is due; a less pedantically accurate Press than ours might have wiped out the whole party on the snout of the glacier.

It was such a hot, sweltering day, and at that height the flogging away of the snow was such exhausting work, that by 3 o'clock, although we had not accomplished a great deal we were of the opinion that it was time to stop. The making of mere steps was not enough. Three or four feet of snow had to be scooped out of the slope to allow room for the body, and a continuous track wide enough for both feet stamped out. We put in pegs to secure the rope as far as we had gone, and then gave the word for retreat. Back at camp at 4 o'clock we found Shipton and Smythe who had re-crossed the Lhakpa La very early that morning on account of the very bad snow conditions on that side. They had watched our performance with interest, not untinged with anxiety, and were glad to see us coming down. Shipton brought a cold back with him, but the party as a whole were now much less troubled with coughs, colds and throats than they had been in April; of course, in view of the change in the weather, this was only to be expected.

At 5 o'clock that evening a snowstorm began which went on for several hours. Until far into the night avalanches could be heard roaring down off the north-east ridge of the mountain opposite, and from the North Peak at our backs. This fall of snow, and the hot muggy morning which followed, gave us something to think about. It was certain that the slopes would be unsafe for two or three days, and with the coming of the north-west wind, which we confidently hoped would presently begin blowing to clear the snow off the mountain, the formation of wind-slab snow might make them dangerous for an indefinite period. This is a snow formation due to wind-driven snow, fallen or falling, being deposited on lee slopes; it is very liable to avalanche and difficult to detect. The avalanche in which Shipton and Wyn Harris were nearly involved in 1936 was thought to be of this type. The uncertain behaviour of the snow on these slopes in other years, once milder weather had set in, was a menacing thought never far from our minds. There was the 1922 disaster when seven porters lost their lives, the narrow escape of 1936, and the more unaccountable happening of 1935, when for no easily ascertainable reason a great avalanche broke away almost across the whole slope. Of this Shipton wrote: 'We were brought up short at the brink of a sudden cut-off which stretched for several hundred yards in either direction. This indicated that an enormous avalanche had recently broken away largely along the line of our ascending tracks. In fact the whole face of the slope had peeled off to a depth of six feet. Very little new snow had been deposited on the slopes and this cannot have had any appreciable effect on the stability of the old snow which we had unanimously agreed seemed perfectly sound.' All that we did know for certain was that we knew very little at all of how the snow on these slopes would behave from day to day.

In these circumstances our thoughts began to turn to the route up the western side of the North Col, which had been warmly recommended by the 1936 party as a useful alternative route if monsoon conditions rendered the east side dangerous as is usually the case. In June that year they had examined the slope from close underneath it, but had thought its ascent inadvisable. However, they considered it definitely safer from avalanche dangers, especially from those arising from the presence of wind-slab. This may well be, but the angle is uniformly steeper, and its surface, unlike that of the eastern side, which is broken up by crevasses and ledges, is a straight slope which is liable to avalanche at any place. As the upshot of a long discussion on 21 May, we decided that Shipton and Smythe should go round, and, if they succeeded in getting on to the Col, should make an attempt. They were to have the use of all the available porters for two days, after which thirteen would return and go up from this side with us if conditions allowed. The plan was not ideal; it would take a week to execute and the contemplated division left both parties numerically weak in porters. But it did promise that one or other party would reach the North Col.

A windy night, followed by a fine cold morning which gave

promise of better weather, made us drop the plan for the moment. Uncertainty whether the western side would 'go', or whether it was any less dangerous than this side, made us reluctant to commit ourselves to it yet. On the 23rd there was no wind. High cirrus clouds were drifting from the north-west, but clouds again began to appear very early over the Rapiu La to the south-east. Lloyd was now in bed with influenza and would have to go down when strong enough to walk. Shipton, Smythe, Oliver and myself, went half-way up to the Col and found the condition of the snow good enough to warrant another start; so on the 24th all of us, except Lloyd who went down to recuperate, made the ascent to the Col with twenty-six porters. Shipton and Smythe, starting soon after 6 o'clock, had finished the remainder of the traverse and were on the top by 9.30 o'clock. Odell and Oliver followed half an hour later, so that when Warren and I arrived with the porters at the beginning of the traverse, soon after 10 o'clock, most of the rope required was fixed ready for use. Warren led the porters while I followed behind nursing a fresh infirmity—a pain in the ribs which for the remainder of the campaign so adversely affected three of the unavoidable functions of the human body on Mount Everest that breathing became uncomfortable, coughing painful, and sneezing agonizing.

So nervous were we about the traverse that at first we sent the men over one by one, but as confidence increased we put five on a rope. The traverse finished along the underlip of a crevasse beyond which there was a very steep snow slope, broken half-way up by another crevasse, leading to the crest of the Col 200 ft. above. This was made safe with ropes which the porters used to haul themselves and their heavy loads. For a heavily laden man the crevasse in the middle was a formidable obstacle, so there Odell spent a busy hour shoving lustily on the behind of each man as he strove to bridge it with his legs. Nearly two hours were spent in climbing the last 300 ft. We dumped the loads about 50 yards north of the spot where we reached the crest, on the site of the 1936 camp. Of this the only indication was the apex of a pyramid tent just showing through the snow. The men burrowed down into it for six or seven feet but all they retrieved for their labour was a rubber mat. We descended on five long ropes and were back at Camp III by 3 o'clock.

Next day Smythe and I went up again with fifteen porters. The importance of an early start had been made so manifest the day before that we left early and reached the dump on the Col by 10 o'clock. The heat was oppressive and the air around us still as we lay outside the tent in which the loads were dumped, dejectedly regarding the snow-covered mountain. Norton's Traverse was white; the snow beneath us deep and soft; on either side, from the Lho La and the Rapiu La, clouds billowed up; and over towards Sikkim a sea of turbulent white cloud lay between us and the distant summit of Kangchenjunga just appearing above it. Even so we were not entirely convinced that the monsoon had begun; like a man marrying for the second time, hope was indeed triumphing over experience. Nevertheless, that feeble spark of hope was very near to extinction. At present there was too much snow on the mountain for an attempt to be made, and with the absence of wind and the presence of occasional fresh falls the snow was increasing instead of diminishing.

Upon our return to Camp III more discussion took place when it was decided that Shipton and Smythe should go down to Rongbuk until the mountain was clearer, and that we others should occupy the North Col in order to examine the snow conditions higher up. It was just possible things were not so bad as they looked. We were still reluctant to commit ourselves to the west side route, but meanwhile there was no sense in keeping more men than necessary at Camp III. No one could say if the conditions ever would change for the better; but it would be as well to find out what the snow was really like higher up, and at any rate the weather was favourable in so far that it was warm and windless enough to make high climbing possible.

Although we knew well, from the experience of earlier parties, that there was little likelihood of a successful climb being made in such conditions, not one of us would have been content to come away without putting that knowledge once more to the test.

Nothing was done next day, except that I, fired by the reports of some of the others who had been to the Rapiu La the previous day, walked over there to enjoy the view down into the Kangshung valley. By starting early enough to reach the pass by 7 o'clock I was barely in time. Already clouds filled the valley to the foot of Chomo Lonzo, while half an hour later the black stone-covered glacier far below was submerged, and the first waves of the rising tide of mist were breaking on the crest where we stood. Like the Lho La on the west side of Everest, the rise to the Rapiu La from the north is very gentle and the fall on the south side terrifically abrupt. By descending a short way on a rope held by Lhakpa Tsering from above I could see something of the fluted, snow-covered south side of the north-east ridge.

At 1 o'clock that day the sun temperature was 117° F. and the air temperature 33° F. A heavy and prolonged fall of snow that afternoon and evening made us revert to our earlier plan; so when Shipton and Smythe started for Rongbuk next day, the 27th, they took with them fourteen porters, intending to return, when they did, by the west side. After lunch Oliver and I walked up to the foot of the North Col slopes; whether it was the lunch or the heat, or the customary effect of an off-day, we felt particularly lethargic. We were disgusted to find that the snow on the moraine and on the glacier was now becoming very rotten, but after a very cursory inspection of the snow at the foot we decided that we would go up next day. I fear we were becoming callous or fatalistic about the avalanche danger, which nevertheless we were never able to forget; but as it turned out events justified this attitude, for after the affair of the 20th nothing ever happened or ever looked like happening. We were particularly careful to avoid the slopes after fresh snow had fallen, but there was always the possibility of a repetition of the 1935 avalanche to keep us on tenterhooks.

The thermometer fell to 10° F. that night. Next morning Odell and Oliver left at 6.30 o'clock, and Warren and I followed about 8 o'clock with thirteen porters. A stifling mist hung over the snow slopes from which the sun beat up in our faces as though from a desert of sand. We always found ourselves afflicted with a terrible thirst on the trips to the North Col, whereas above we found no real discomfort. Thermos flasks were always in great request by the time the snow boss below the traverse was reached. The heat, and its possible effect on the snow, induced us to treat this with the utmost respect. We crossed it one by one, securing each man on a rope over the worst bit. We were up soon after midday and proceeded to pitch the dome tent which we had now brought up from Camp III. It comprised two loads of 40 lb., but it was worth having up because all the Europeans could live in it together in comfort. Eight men were sent down in charge of Angtharkay with orders to come up next day with more loads if no snow fell. That afternoon, however, it snowed until 8 o'clock at night. Brewing tea took 40 minutes, but it was satisfactory to find the stoves burning well and the specially fitted pumps, which we were using for the first time, very effective. On the 29th we lay at earth. Nearly a foot of snow had fallen overnight and there was wind and more snow that afternoon. There could be little doubt left now that the monsoon was definitely established—had been probably for the last three weeks. Conversation in the tent was of nothing but sardines and tinned fruit; the atmosphere seemed almost defeatist. It was surprising to be told that the temperature in the tent fell that night to 17° F.; judging from the state of the snow the outside temperature must have been hardly as low. I was out early on the morning of the 30th to rouse the Sherpas who usually take longer to regain full consciousness than most people owing to the overpoweringly poisonous atmosphere which they love to encourage inside their tents by sealing up any crevices suspected of letting in dangerously fresh air. Having broken a pricker in their Primus stove in an endeavour to expedite matters, and set a guard to watch and report any signs of movement down at Camp III—we did not want them to come up—I returned to the tent to brew a handsome dish of scrambled eggs; what the Sherpas expressively call 'rumble-tumble'. Upon this we all became pretty active and by 8 o'clock were in a sufficiently robust frame of mind to contemplate the notion of starting with something approaching Christian resignation. It was then that the look-out reported Angtharkay's party moving up the glacier from Camp III, some 1,500 ft. below us and a mile or more distant. However, a prolonged bellow from all hands on the Col was heard by them, interpreted correctly, and acted upon with almost indecent haste. They returned to camp.

Of our party, Oliver was to examine from the end of a long rope the state of the snow on the western side, while the rest of us went up the north ridge in the direction of Camp V as far as our enthusiasm or energy would carry us. The snow was knee-deep as we steered a careful course amongst the crevasses at the foot of the ridge, and began plodding up the easy slope. Warren tried out the 'closed' type apparatus with unlooked-for results. Instead of making him skip upon the mountains like a young ram it seemed bent, first on bringing him to a standstill, and then on suffocating him. He did not wear it very long. Whether there was something wrong with the works or whether there is something inherently vicious about the design has yet to be discovered. Its fellow was already definitely broken but this one appeared to be functioning correctly.

In front we had the tireless Tensing to make things easy for us until we came to a steeper slope where the snow seemed very insecurely poised. Here I went on ahead, bringing up Tensing with the help of another rope which Odell had been carrying, for I had to go on for two rope-lengths before finding a secure stance. Then as we could not get this rope back to Odell he waited there for Warren and Oliver while I pushed on slowly with Tensing. At 1 o'clock we were still on the long snow-bed which permanently covers the lower part of the north ridge to a height of nearly 25,000 ft. We were just below the bulge where the angle eases off, at a height of about 24,500 ft., and there we sat down feeling remarkably little inclination to go on. Tensing complained that his feet were cold—they had probably got wet. But it was a pleasant day—I myself wore neither wind-proofs nor gloves—and though most of the country below was hidden by a veil the clouds themselves formed a wonderful and constantly changing picture. Oliver was coming on from below, but I signalled him to stop, for we had seen enough to realize that little would be gained by occupying Camp V in those conditions. We went down; and that evening, when more snow fell, decided to abandon Camp IV for the present. A change for the better might come sometime, but even if no change came I was eager to return to give the upper part of the mountain a trial.

As we were anxious to start early next morning while the snow was in good condition, we decided, rather unwisely, as it proved, to postpone breakfast until we got down to Camp III. In mountaineering, as in war, it seldom pays to defer a meal if an opportunity for eating offers; at the best the next opportunity may be long in coming, at the worst it may never come again. A suggestion that two of us should go down by the west side to acquire first-hand knowledge of the route, and possibly to meet Shipton's party, was coldly received on account of Oliver's adverse report on the state of the snow there. We struck the dome tent before leaving to prevent it blowing away, but left the pyramid and the small Meade tents standing with the stores piled inside. The descent took longer than expected. At first the snow was good, but on the lower half the thinness and looseness of the snow, with the presence of ice beneath, imposed the necessity for much step-cutting and of constant vigilance to prevent the porters taking liberties, as those without much ice experience are apt to do. We were feeling pretty limp when we got back to Camp III about noon, not alone from hunger but also from heat. Snow began falling at 1 o'clock and continued until 5 o'clock. The joy occasioned by the arrival of the mail which might have helped one to forget the weather was offset for me by the news that Rs. 800 had been stolen from the cash-box at Rongbuk.

15. En route from Base Camp to Camp I (p. 50)

16*a*. Eric Shipton near Camp I (p. 51)

16*b*. Alpine choughs on a glacier-table at 20,000 ft. (p. 51)

CHAPTER VIII

THE WESTERN APPROACH AND DEFEAT

The attempt and not the deed confounds us.

SHAKESPEARE, *Macbeth* II. ii. 12.

A DISTINCT change in the weather occurred on 1 June. The morning was dull and by midday the sky was covered with low clouds driven swiftly before a strong west wind. Taking only our bedding and some tents we walked down to Camp I where we expected to meet the others or at any rate to hear news of them. Amongst the ice pinnacles bordering the glacier trough the warm weather had wrought great changes; everywhere water ran or formed in deep pools. Close to Camp I the glacier stream, which before ran silently under a deep covering of frozen snow, was now a grey flood crossable only at certain places. A Sherpa, whom we had sent on ahead, reached the camp in time to find some of Shipton's porters engaged in relaying loads up to Lake Camp where his party now was. A chit having been sent up telling them to wait, Oliver and I next morning walked up to discuss plans.

A glance at the map will show that Lake Camp is only about 1½ miles from Camp I; the approach to it is across the East Rongbuk valley, below the glacier snout, and then up the moraine shelf on the right bank of the main Rongbuk glacier. Camp I is as good a camp as one could expect at 18,000 ft., but Lake Camp is better. Of course there is no fuel, but it is warm and sheltered, for it lies tucked away between a high moraine bank and the bounding wall of the valley; a stream of clear water meanders through a grassy lawn to empty itself into a little lake which is pleasant to look at if not to bathe in. Because of the sheltering slopes the morning sun reaches it rather late in the day, a disadvantage which is offset by the fact that it is open to the sun until a late hour in the afternoon. No one, I imagine, who visits this camp is likely to grow tired of looking at grass, for as the shadow of a great rock and water in a thirsty land so is grass to the sojourner among ice. But should he do so he has only to climb the moraine bank to see Everest with its stupendous north-west ridge sweeping down to the Lho La, the fascinating pinnacles of the main Rongbuk glacier, and the striking group of mountains, Pumori and the Lingtren group, lying between the West Rongbuk and the upper main Rongbuk glacier.

Shipton, Smythe and Lloyd, who had now recovered, were on their way up to the west side of the Col; but having heard that we had come down became uncertain what to do and were toying with the idea of making an excursion up the West Rongbuk glacier. However, this change in the weather kindled fresh hope in us all. By stepping a few yards from the camp we could see the north face of the mountain, and, through breaks in the flying scud, snow being whirled off it in the most encouraging manner. If this continued we might yet accomplish something, but while the wind promised to clear the mountain it also threatened to form wind-slab snow on the east or lee side of the Col. Bowing to Smythe's repeated warnings, we decided to abandon that side altogether and to concentrate on an approach from the west. I attached myself to Shipton's party in order to make up two climbing parties of two, and we brought up our porter strength to seventeen by sending for two good men from Camp I. The remaining fourteen porters were left to bring down some necessary loads from Camp III, and then, with the three Europeans, were to follow up as quickly as possible.

The wind having blown for 48 hours dropped on the 3rd, a very fine, almost cloudless day, with very little wind at all except high up. The march to the next camp is a rough one; at first over moraine boulders and later along a trough on the right bank of the glacier. The camp is close to the point where the short glacier leading to the west side of the Col joins the main glacier. Tents were pitched on the glacier itself, the coolness of the ice being tempered by the layer of stones which covered it. Some debris left by the 1936 expedition, including a wireless aerial, still lay about—evidently the Tibetans have not yet heard of this departure from the usual route. This place was called North Face Camp but perhaps Corner Camp would be a better name. It was so mild that afternoon that we sat about on the moraine boulders talking until after 5 o'clock; the ancient philosophers sat on stone seats expounding their systems, and no doubt our topics—the absurd height of Mount Everest and the evils of processed foods—were so congenial that we too were indifferent to comfort.

Some snow fell in the night, and the morning was cloudy when we left at 9.30 o'clock. Turning the corner we proceeded, roped, up the glacier, steering a course close to the slopes of the North Peak above us on our left. The snow was good. The clear view we had of the north-west ridge of the mountain again impressed us with the difficulty a party would have in setting foot on it. When we had made about a mile up the glacier an icefall compelled us to bear to the left still closer under the North Peak, but as we changed direction mist came down to obscure the route and at the same time we noticed a lone figure hurrying after us up the glacier. The figure proved to be Lhakpa Tsering, who, as he had not been able to join us yesterday, was now coming straight through from Lake Camp. Leaving some men here whom he could join on the rope, for there were a number of crevasses about, we pushed on through the mist with eyes and ears cocked for anything which might fall from the North Peak so unpleasantly close above us.

By now the snow was rapidly deteriorating, but, as the many big crevasses were adequately bridged, soon after 1 o'clock saw us camped in the middle of the wide snow shelf above the icefall, two or three hundred yards from the foot of the slope leading to the Col. On account of possible avalanches it was advisable to keep as far away as possible from the snow-slopes surrounding us on three sides. The height of this West Side Camp must be approximately the same as that of Camp III, about 21,500 ft. The afternoon was fine and hot; on the Col snow was still being blown about, and far above us on our right the rocks of the Yellow Band looked deceptively free from snow. The outlook, in fact, was encouraging.

We breakfasted at 5 o'clock and left at 7 o'clock of a bitter cold morning, already chilled to the bone by the inevitable waiting for loads to be adjusted and frozen ropes disentangled. The sky had a curiously dull glassy look against which, to the west, the beautiful Pumori (the 'Daughter Peak' of Mallory) appeared flat as if painted on grey cloth. Smythe and Lloyd went ahead to make the track, the rest of us following on three ropes. As we made our way to the foot of the slope the most phlegmatic might have remarked on the fact that we were walking over the debris cone of the father and mother of all avalanches, the tip of which reached nearly to the camp. This had apparently fallen a few days before—possibly on the day on which I had suggested two of us should descend this way from the Col. One result of this fall was that the first five hundred feet of our route now lay up bare ice, thus entailing much hard work for the leaders step-cutting; and in order to reach snow which was still in place, and which at that early hour, if our pious hopes were fulfilled, might possibly remain in place, a long traverse had to be cut across the ice, in crossing which it was impossible adequately to safeguard the porters. Two of us were roped to the porters, but this was done in the hope of inspiring confidence and not in the expectation of checking a slip. All that one could do really was to urge caution and pray that no slip occurred. Having reached the snow we were faced with about 800 ft. of steep slope; but the snow was of doubtful integrity and of a consistency that gave us plenty of hard work. The appearance of the sun over the Col, warned us of the need for haste, so we pressed on with all the speed we could muster. When we reached the top at 11 o'clock the sun, feeble though it was, had been on the slope for an hour and from its effect on the snow I felt it was high time for us to be off it. It was fortunate that the sun that morning was but a pale reflection of its usual self, for it peered wanly through the glassy sky surrounded by a double halo. Surprisingly enough no abnormal weather followed these alarming portents. Some snow had accumulated round the tents which we had abandoned on 30 May, but nothing had blown away. We re-pitched the dome tent, and four Europeans and sixteen Sherpas (one was left behind sick) were once more in occupation of Camp IV.

6 June dawned fine. As one of the porters had gone sick we had only fifteen available for load carrying, but this was just sufficient if they carried between 25 and 30 lb. Some time was spent making up loads which had to contain all that was required for Camps V and VI; that is to say, tents for two Europeans and seven Sherpas at Camp V, and for two Europeans at Camp VI; sleeping-bags and food for three days for two Europeans and seven Sherpas at Camp V, and sleeping-bags and three days' food for two Europeans at Camp VI. We took two pyramid tents to Camp V. They make heavy and awkward loads but would enable a party to sit out any bad weather in reasonable comfort; a similar type is used in the Arctic where their great advantage is that the four poles have merely to

be brought together to enable the tent to be placed on a sledge ready for travelling. Pitching is equally quick and simple, but this is not possible when the tent has to be carried by a man, for then the poles have to be taken out and disjointed.

We finally got away at 10 o'clock to make good progress up board-hard snow. A week ago we had sunk to our knees in the snow of this north ridge, but now it was so hard that a strong kick was needed to make a nick for the edge of the boot. The question of the behaviour of snow at high altitudes is difficult. For example, this hard snow extending up to 25,000 ft. gave us some reason to expect fair snow conditions higher up, but above Camp VI it was loose and soft and apparently quite unaffected by the wind. A variety of factors have to be considered besides the mere force and direction of the wind which may of course vary greatly in different places and at different heights. Mr Seligman, who is an authority on Alpine snow conditions, believes that if some evaporation of snow cannot take place no packing can occur. That is to say, the wind must be dry. Further, the lower the temperature the less easily does snow evaporate so that it is possible that above a certain height temperatures are too low for any evaporation to take place.

Lloyd was wearing the 'open' type oxygen apparatus in which he went well once he had mastered the technique of breathing. Not that he climbed any faster, but perhaps he did so more easily. The hard snow-bed which afforded such excellent going comes to an end at something below 25,000 ft. Here there is a momentary easing of the slope where a party can sit comfortably while it summons up the little energy remaining for the next 800 ft. of mixed scree, rock and snow. Above this two of the men, who were feeling the height, struggled on for only another 100 ft. before giving in altogether. They had to be left where they were together with their loads. Tensing was going very strong, but none of the others seemed at all happy. When still some 300 ft. below the site for Camp V (25,800 ft.) a sudden snowstorm sapped their resolution so much that there was talk of dumping the loads and going down. By now Smythe and Lloyd were nearly up; they must have wondered what Shipton, myself, and the porters, were doing, strung out over the ridge in various attitudes of despair and dejection hurling remarks at one another. In the end better feelings prevailed. All struggled on, and by 4 o'clock reached the fairly commodious snow platform of the Camp V site.

Unluckily one of the two abandoned loads was the second pyramid tent. Consequently, the sole accommodation for two Europeans and seven porters was a small Meade tent (later to go to Camp VI) for the former, and one pyramid which at best would take five Sherpas. The prospect of sleeping seven in a tent made to hold four was bleak, but that of going down and returning next day was worse. Lloyd and I started down with six porters at 4.15 o'clock, but we had not gone far before shouts floated down to us from above. I could not understand what was said, but apparently it was a request to us to bring up the other tent which had been dumped six or seven hundred feet below. The only possible reply to this was the Sherpa equivalent for 'Sez you'; for I do not think that any of our party, had they been willing, were capable of doing it. We carried on, picked up the two sick men, and continued the descent very slowly; for the men were too tired to be hurried and called frequently and successfully for halts by the simple expedient of sitting down. The snow, too, was now very soft. Looking back at one of these halting places I saw, descending from Camp V, two men whom our Sherpas recognized as Pasang and Tensing coming down to retrieve the abandoned loads. To descend and ascend with loads another seven hundred feet, on top of the toil they had already endured, was a remarkable example of unwearying strength and vitality gallantly and unselfishly applied. We reached Camp IV at 6.15 o'clock, pleased at having at length established Camp V. For my part I was very hopeful that something might yet be done. The last entry in my diary for that day runs: 'Frank and Eric going well—think they may do it', which showed how little I knew.

Camp V was established on 6 June; the party's activities in the following days may best be told in Shipton's own words.

On 7 June a heavy wind was blowing from the east; this prevented our advance up the ridge. The weather, however, had been fine for a week, and as there had been a lot of wind we hoped that the snow would be coming off the mountain. The following day was calm and fine. We started at 8 o'clock. The whole party was fit and full of hope that we were going to be granted a chance for an attempt at the summit, which had been denied us for so long. The upper part of the mountain was very white. It had always been presumed that when it presented such an appearance there was little chance of success. But no one had ever climbed far above the North Col during the monsoon, and this idea had been founded on pure conjecture. Conditions on the ridge as far up as Camp V had led us to hope that with the recent fine weather and cold winds the snow on the upper slabs might have consolidated; for now it was clear that at this time of year (on account of the increased humidity) no amount of sun and wind would remove it.

We had not gone far before we found that our hopes were vain. The rocks were deeply covered in snow, which, unlike that below Camp V, showed no tendency to consolidate and was as soft and powdery as it had been when it had fallen about ten days before. The ridge which in 1933 had not caused us the slightest trouble now demanded a lot of very hard work. It was almost unbelievable that such a change could take place on such simple ground. There was one small step that both Smythe and I failed to climb, and we wasted a considerable time making a way round it. It was hard work, too, for the porters, and our progress was lamentably slow. It was 1 o'clock before we reached the site of Norton's and Somervell's old camp at 26,800 ft. The porters worked splendidly and without any complaint. They were determined to put us in the very best position possible from which to make our attempt, and would not listen to any suggestion that they might have difficulty in getting back before nightfall. Previously it has always been rather a question of driving these men to extreme altitudes; now the position was almost reversed. I do not think future expeditions need worry about the establishing of their higher camps provided they choose the best men. Pasang was not well and his comrades went back to help him with his load more than once. Ongdi, too, showed signs of great exhaustion.

At the top of the north-east ridge, we reached, at 4.15 o'clock, a gentle scree slope below the Yellow Band. Here we pitched our tent at an altitude of 27,200 ft. I have never seen the Sherpas so tired, and they must have had a hard struggle to get back to Camp V before dark.

The weather was fine, and the sunset over hundreds of miles of monsoon clouds far below was magnificent. But all we wanted to do was to lie quietly down in the drowsy condition which seems to be a permanent state at great altitudes. It was a big effort to cook and eat any supper, and all we could manage that night was a cup of cocoa and a little glucose. I had brought a small book with me against the possibility of a sleepless night. But the meaning of the words kept becoming confused with a half-dream, as when one is reading in bed late at night before going to sleep.

We started cooking breakfast at 3.50 o'clock, and started before the sun had reached the slabs of the Yellow Band. But we were surprised to find the cold was intense. Very soon we had lost all feeling in hands and feet, and it was obvious we were in serious danger of frost-bite. We returned to the tent and waited until the sun had arrived, and then made a second start. Norton's route below the Yellow Band was quite out of the question for there was an enormous deposit of snow on the gently sloping ground. Also conditions in the couloir were obviously hopeless. Our plan was to try to make a diagonal traverse up to the ridge which we hoped to reach just before the First Step. At best it was a forlorn hope, for the ridge in any condition must be a tough obstacle, and it now looked really villainous. The only chance lay in the remote possibility that some unexpected effect of wind and sun at these little-explored altitudes had produced firm snow on the steep slabs and on the ridge.

We started flogging our way up the steep ground, through powder snow, into which we sank up to our hips. An hour's exhausting work yielded little more than a rope's length of progress, even on the easy beginning on the slabs. We went on until, on the steeper ground, we were in obvious danger of being swept off the rocks by a snow avalanche. Then we returned, completely convinced of the hopelessness of the task. It was bitterly disappointing, as we were both far fitter at these altitudes than we had been in 1933, and the glittering summit looked tauntingly near.

There can be no doubt that one day someone will reach the top of Everest, and probably he will reach it quite easily, but to do so he must have good conditions and fine weather, a combination which we now realize is much more rare than had been supposed by the pioneers on the mountain. It is difficult to give the layman much idea of the actual physical difficulties of the last 2,000 ft. of Everest. The Alpine mountaineer can visualize them when he is told that the slabs which we are trying to climb are very similar to those on the Tiefenmatten face of the Matterhorn, and he will know that though these slabs are easy enough when clear of ice and snow they can be desperately difficult when covered in deep powder snow. He should also remember that a climber on the upper part of Everest is like a sick man climbing in a dream.

So much for the first attempt made this year, if indeed attempt is the right word; for neither of the parties which started out from Camp VI this year was under any illusions regarding the possibility of reaching the summit. A more accurate description of the final stages of this year's expedition would be that a brief inspection of the conditions above Camp VI satisfied both parties that reaching the summit was then an impossibility. And, as Shipton says, this was the more disappointing because both he and Smythe were feeling strong enough to justify some confidence in the result had conditions been favourable. This point needs stressing in view of the argument sometimes brought forward that small party methods impose too much strain on those taking part.

17*a*. The Kharta valley: Arun gorge in distance (p. 62)

17*b*. Rest camp in Kharta valley (p. 62)

18*a*. Negotiating the descent from Lhakpa La, 22,200 ft. (p. 61)

18*b*. Gorge of the Arun river (p. 62)

CHAPTER IX

ATTEMPT TO REACH SUMMIT RIDGE

How pitiable is he who cannot excuse himself.

DOWN at Camp IV on 7 June we were almost as inactive as the party at Camp V—but with less reason, for here there was no wind and the day was fine. The temperature at night outside the tent had been 5° F. We rose late, and after a leisurely breakfast Lloyd put on the 'closed' type oxygen apparatus and started to walk towards the foot of the north ridge. I watched with interest and soon saw that all was not well. When he was about 200 yards from camp he sat down, took it off, and came back. The feelings of suffocation which he experienced were exactly similar to Warren's. I put the thing on for myself for a few minutes and executed a light fandango in the snow with such remarkable feelings of sprightliness that I resolved to give it a proper trial. I regret to say that this resolution, like so many others made at high altitudes, was not kept; but there is no reason to believe that my experience would have differed from that of Warren or Lloyd who had both given that apparatus a fair trial. They could find no mechanical defects at the time, but whatever faults may have since been detected, and whether or no they can be remedied, I think the most obvious lesson to be learnt is that the only trials and experiments of any value at all are those carried out by mountaineers themselves at heights of over 23,000 ft., but not necessarily on Mount Everest.

We were up very early on 8 June in order to take three sick men down to the West Side Camp; they were not exactly sick but were incapable of acclimatizing sufficiently to go higher. Therefore they were better out of the way. As Lloyd and I had to return it was important to start early in order to have safe snow conditions on the way back. The men's own sleeping-bags were down at the West Side Camp; for use on the Col we had a common stock of high altitude sleeping-bags for every one which were of course much warmer than the ordinary bags. When these three worthies heard that no sleeping-bags were to be taken down they became recalcitrant and refused to go, but as it was too early in the morning, and too cold, for argument, we tied them on to the rope and started off. We were fortunate that the temperature during the night had fallen to 5° F., consequently the snow was in such excellent condition—firm but not too hard—that we descended the snow slope in half an hour. While I belayed the party so far as any satisfactory belay was possible on the ice, Lloyd recut the steps across the traverse. Once across that we left the men to look after themselves for the short and safe remaining distance down to the tent left there, while we returned to the Col reaching it at 9 o'clock. Although the sun had not yet touched the slopes, the crust which had supported us going down now began to break under our feet. For the rest of the day a mist which enveloped the Col prevented us from seeing what the Camp V party were doing.

Next day, 9 June, saw us moving up to Camp V in accordance with a prearranged programme. We took with us six porters lightly laden with a little extra food and five oxygen cylinders. Lloyd again wore the 'open' oxygen apparatus while I was without. Thus we constituted rather a hybrid party but such a party might function quite well for an attempt on the summit. The reason that I was without was not solely one of high principles or an intolerant scorn for the use of oxygen apparatus, but because if we were to use it all the way, as Lloyd intended, then there would not be enough cylinders to supply the two of us.

As we began the ascent of the north ridge we saw the rear party —Odell, Warren, Oliver—accompanied by two Sherpas coming up from West Side Camp. They had crossed the traverse, but as we could not afford the time to wait and exchange news we continued on our way. A little later we met the seven men coming down from Camp V, very tired; their names were: Ongdi Nurbu ('Ashang'), Pasang Bhotia, Rinsing, Tensing, Lhakpa Tsering, Da Tsering, Lobsang. About this time Shipton and Smythe must have been leaving Camp VI, so altogether the mountain presented such a scene of activity that morning that it reminded one of Snowdon on a Bank Holiday. On reaching the top of the snow slope one of our men, whose cough was troubling him, dropped out, but the rest of us reached Camp V at 3 o'clock almost simultaneously with the arrival of Shipton and Smythe from above. Their report of conditions up there effectually quenched any hopes we may have entertained of reaching the summit. A valuable plan was to investigate the possibilities of the north-east or summit ridge, and, if we could reach it, to have a look especially at the Second Step. In any case, with such snow conditions, the ridge was undoubtedly a safer place than the slabs.

After some tea and talk they went down, taking with them three of our porters. Kusang Namgyal and Phur Tempa stayed with us; the former needed no prompting, electing to stay as of right and privilege, but some persuasion was required before one of the other four volunteered. It is of course a matter for wonder, no less than thankfulness, how much these men will do and how far they will go with, one imagines, few of the incentives which act as a spur to us. There is, as Shipton remarks in the passage I have quoted, no longer any need to drive them to go high if care is taken to pick only the best. It was not always thus. Amongst other factors which have brought about this change, and to which we are chiefly indebted, are the care, sympathy and mountaineering skill, with which the porters of earlier Everest expeditions have been handled. The value of the confidence which a leader and his party now derive from the knowledge that the porters will carry a camp as high as need be cannot be over-estimated.

We slept little if at all that night. We were warm enough, in spite of having only the tent floor between our sleeping-bags and the snow, but from sunset until four in the morning it blew so hard that the noise made by the flapping of the double-skinned pyramid tent kept us awake. The night temperature was $-1°$ F. We got away soon after 8 o'clock with Kusang and Phur Tempa carrying between them two oxygen cylinders and the little extra kit and food we needed. Camp VI was already provisioned and equipped with sleeping gear. With snow lying everywhere the climbing for the first thousand feet was not easy, and the tracks of the first party had for the most part been filled in by the wind. Lloyd went ahead making the track while I followed roped to the two porters. Although it was naturally harder work making the route, Lloyd reached the camp some half an hour before us, and, by the time we arrived, had re-erected the small Meade tent which the others had struck for the sake of safety. As was to be expected the higher we went the more benefit he felt from the oxygen. On the other hand,

I was roped to the two laden porters, and though at this distance of time I like to think I was accommodating my pace to theirs I should not like to have to take my oath upon it. For the short distance we went next day he again went better than I did, but perhaps under the circumstances that is not such a valuable testimonial for the oxygen apparatus as it might seem. It is conceivable that a better man, or a man who had had less sickness during the previous month, might have gone as well as or better than Lloyd, who, it must be remembered, was carrying a load of 25 lb. I am inclined to think that any benefit likely to be obtained from the use of oxygen is cancelled by the weight of the apparatus. But this may not always be the case—a very much lighter apparatus is probably only a matter of time. It is well known that the effect of great altitudes is to sap not only the powers of the body but also of the mind. To say that the resolution of every man who goes high is thereby weakened to some extent would be too sweeping, but the will, or even the urge, to ascend and to overcome the difficulties standing in his way, which is the instinctive feeling of every mountaineer, is less strong than at lower levels. What I did rather hope and expect, therefore, was that the revivifying effect of oxygen might be sufficient to overcome this disquieting tendency, and that a few whiffs of oxygen would boost Lloyd up those rocks which next day so easily defeated us—and of course would enable him to pull me up too. But, as will be seen, there was no such effect; oxygenated and unoxygenated man acquiesced tamely in defeat.

Shortly before reaching the 1924 camp site (26,800 ft.) we untied the rope as here there was little likelihood of a slip having any serious consequences; but it was needed again on the smooth slabs just below the tent, which a covering of snow made very awkward indeed. At the top of these slabs the angle eased off, and a gentle slope of small scree, almost free from snow, continued for nearly a hundred yards up to the foot of a steep rock wall. For thirty or forty feet this wall was very steep—steep enough that is to necessitate the use of the hands—until the angle eased off where broken slabby rock led to the north-east shoulder (27,500 ft.) and the summit ridge, about 300 ft. above our camp. Fifty yards to the left (east) of the scree patch the north ridge fell away steeply to the big snow couloir which has its origin almost directly below the shoulder.

On the right (west) at the same level a wide bed of snow ran up to the foot of the heavily snow-covered rocks of the Yellow Band, the wide band of light-coloured rock running almost horizontally across the face of the mountain. The whole face looked steeper and more formidable than I had imagined. We had placed Camp VI two or three hundred yards to the east of and about 200 ft. below the 1933 camp site owing to the impossibility of getting there. Ours was less advantageously placed for making an attempt, but otherwise it is probably the best site on the mountain; reasonably flat, and, even at this time, free from snow except for a little hard patch close to the tent from which we drew our supply for cooking.

We were all up by 12.30 o'clock. Having sent the men down we collected snow for cooking and turned in, for the wind was already rising. It blew all afternoon and continued for most of the night, so that again we slept little. For supper that evening we each had the best part of a pint mug of hot pemmican soup which we swallowed with equanimity if not with gusto; nor did I think it was this, and not the wind, which accounted for the rather miserable sleepless night which followed. To drink a cup of pemmican soup is very well, and to do it at that height is indeed a triumph of mind over matter; but the whole cupful does not amount to more than 4 oz. of food. This was probably more than half of our total food intake for that day which again was only a quarter of the amount considered necessary for men doing hard work. Pemmican and sugar have high calorific values—the former the highest of any food—but to provide the requisite number of calories, supposing only these two foods were taken, one would have to consume ½ lb. of pemmican followed by 1 lb. of sugar which would at any rate help to keep it down. Perhaps this could be accomplished easily enough in the course of a day; but how many pounds of caviare, quails in aspic, chicken essence, sweet biscuits, jam, dried fruit, tinned fruit, pickles, and the other things beloved of those who believe this high altitude food problem could be solved by the tempting of the appetite, would be required to do the same? In many counsellors there may be wisdom, but in many foods there is neither sense, nourishment, nor digestibility. There are several foods still to be tried out, but I believe the fact will have to be faced that at these altitudes it is impossible to eat very much and that one will have to be satisfied with eating as much as one can of some food of high value, even if not very palatable, rather than pecking at kickshaws. In spite of having been inhaling oxygen for most of the day Lloyd had no more appetite than I had, but perhaps he ought to have taken alternate sucks at his mug of pemmican and his tube of oxygen. Proverbially difficult though it is to blow and swallow at the same time, no doubt in the near future some genius will produce an apparatus which will make possible the assimilation of oxygen and food simultaneously.

We rose early on the 11th; not so much because we were panting to be off, like hounds straining at the leash, but because being up would be less wretched than trying to sleep. To say that we rose conveys a wrong impression; we did nothing so violent, but merely gave up the pretence of trying to sleep by assuming a slightly less recumbent position. Then one of us had to take a more extreme step—sitting up, and reaching out for the stove; and for the saucepan full of snow, waiting in readiness at the other end of the tent. Once the stove was lit an irrevocable step had been taken, for it had to be tended. But once lit it burnt well considering that it was labouring under the same difficulty as ourselves, anoxaemia —not so well, however, that one could afford to leave it to its own devices and pretend to go to sleep again. Occasionally it would splutter, which was the signal for going out altogether or for a long tongue of flame to lick the roof of the tent playfully—an emergency calling for some deft work with a pricker and a match until it was burning normally again. In the course of half an hour or so the lifting of the saucepan lid reveals no merrily bubbling water, but a murky pool of slush or half-melted snow, its surface coated with the remains of last night's pemmican. Feelings of impatience for a hot drink and thankfulness for further respite in bed are mingled equally. If patience is bitter, its fruit is sweet. Presently the water bubbles feebly and breakfast is served—a mug of tea, not completely valueless as it contains a good quarter pound of sugar, a few biscuits, and possibly a fig. None of us has been able to face porridge above Camp III, nor is there time for making both porridge and tea.

By morning the gale of the night had died away, but it was not until 8 o'clock that we considered it warm enough to make a start.

(All times given, by the way, are relative and not absolute. It was 8 o'clock by my watch, but by the sun it might have been 7 or 9 o'clock.) While waiting we dressed by putting on wind-proofs, and boots which had been kept more or less unfrozen in our sleeping-bags. The sun was still below the ridge but the morning was fine and calm except for what appeared to be a gentle zephyr from the west. In reality it may have been blowing hard for I suppose if the atmospheric pressure is only one-third of normal, wind strength is also reduced. On our arrival the previous afternoon I cannot say that the rock wall which we proposed climbing as the most direct way to the summit ridge had made a very good impression. Like boxers confidently announcing their victory on the eve of a fight we told each other it would 'go', comforting ourselves privately with the thought that rocks sometimes look worse than closer acquaintance proves them to be. But now, in the cold light of morning, as they looked still less prepossessing we decided not to waste time but to turn the wall on the right where it merged into the easier angle of the face, and where a shallow depression filled with snow led diagonally upwards to the summit ridge. As we moved slowly up the scree towards the right-hand end of the wall I kept changing my axe from one hand to the other thinking it was that which was making them so cold. But before we had been going ten minutes they were numb and I then began to realize that the gentle zephyr from the west was about the coldest blast of animosity I had ever encountered. I mentioned the state of my hands to Lloyd who replied that his feet were feeling very much the same. We returned to the tent to wait until it was warmer.

We made a second brew of tea and started again about 10 o'clock by which time the sun had cleared the ridge, although it was not blazing with the extraordinary effulgence we should have welcomed. In fact at these heights the only power which the sun seems capable of exerting is that of producing snow-blindness. It was still very cold, but bearable. We skirted the snow lying piled at the foot of the wall and took a few steps along our proposed route, where Lloyd, who was in front, sank thigh-deep into the snow. I believe it was somewhere about here that Shipton and Smythe had tried. Without more ado we returned to the rocks. There seemed to be three or four possible ways up, but first we tried my favoured line of which Lloyd did not think very highly. It was one of those places which look so easy but which, through an absence of anything to lay hold of, is not. I did not get very far. A similar place was tried with like result and then we moved off to the left to see if there was any way round. This brought us to the extreme edge of the north ridge where it drops steeply to the gully coming down from the north-east shoulder. There was no way for us there. Retracing our steps along the foot of the wall Lloyd had a shot at my place which he now thought was our best chance; but he too failed. It was not really difficult; at least looking back at it now from the security of an armchair, that is my impression, but the smooth, outward-sloping rocks, covered in part by snow, very easily withstood our half-hearted efforts. I then started up another place which I think would have 'gone', although the first step did require a 'shoulder'. Very inopportunely, while I was examining this our last hope, there was a hail from below and we saw Angtharkay, presently to be followed by Nukku, topping the slabs just below the tent. I had left word for him to come up with the oxygen load abandoned by the porter who had failed to reach Camp V with us. We wanted to have a word with him, and of course to go down to the tent was a direct invitation to go down altogether—a course which I am sorry to say was followed without any demur.

It will be a lasting regret that we never even reached the summit ridge, but I think the information we would have brought back had we reached it would have been mostly of negative value. From the point we were trying to reach close to the north-east shoulder, the Second Step is about 1,200 yd. distant; the summit itself is a mile away, and 1,500 ft. higher. The ridge, on which there was plenty of snow, did not look easy, while the Second Step looked really formidable; so much so that the only chance seemed to lie in the possibility of making a turning movement on the south face, which of course we could not see. The only reason for preferring the ridge route to Norton's Traverse would seem to be when there is snow about; but since under such conditions the ridge itself is not easily attainable little remains to be said for it as an alternative route.

We descended to Camp V in a storm of snow and wind which made the finding of the best route a matter of difficulty. Kusang and Phur Tempa were still in residence. After some tea and an

19. Approach to eastern face of North Col (p. 67)

20a. The Camp on North Col, North Peak in background (pp. 74 and 80)

20b. Chang Tze (North Peak) and North Col (below) from Camp V, *25,600* ft. (p. 82)

hour's rest we started again at 4 o'clock for the Col. The Sherpas were roped together, while we went ahead making a track for them down the snow; but they went so slowly that we had constantly to wait for them. The storm had blown itself out and the evening was now calm and fine, so near the bottom we pushed on ahead leaving them to follow at their leisure. Amongst the crevasses at the foot of the ridge, where the storm had obliterated all old tracks, we had a discussion, about the right route, which threatened to be interminable until Lloyd settled the matter, or at any rate pointed out the wrong route, by falling into one; thus bringing an inglorious day to its appropriate conclusion. As we were unroped at the time, this slight mishap will possibly evoke neither surprise nor sympathy. In response to my inquiries a muffled cry from below assured me that he was unhurt and had not fallen very far; but as nothing could be done until the porters arrived with the rope, I had to leave the victim down there for a good ten minutes—possibly penitent, certainly cold. It should never happen, but if one does fall into a crevasse in free unfettered fashion (I speak from experience), as one does if the rope is not being worn, it is a question which feeling predominates—surprise, fear, or disgust at having been such an ass. I could see the Sherpas up the ridge and they could see me, but they neither heard my shouts nor took any notice of my gesticulations except to sit down once again and ponder at this new form of madness. At length one of them, who was possibly a better arithmetician than the others, must have totted up the number of Europeans who had left Camp V that afternoon and discovered that now there was one short. Down they came, and Lloyd was hauled out none the worse.

Only Odell, Oliver and seven porters remained at Camp IV as the others had gone down that morning by the old route to Camp III. Warren was obliged to go too in order to look after Ongdi who on his return from Camp VI had suddenly developed pneumonia; at least the symptoms pointed to pneumonia, but his recovery was so speedy that it may not have been. They went down by the old route because for a sick man that was the easier. Another Camp VI man, Pasang Bhotia, was lying there alone in a tent, sick. The first report was that he had gone mad for he was unable to articulate, but it soon became clear that his right side was completely paralysed and he was therefore incapable of movement. They had attempted to take him down that morning, but two sick men in one party were too many. The other Sherpas seemed rather to wish to shun him than to assist him in his piteous plight; he could neither dress himself, put on his boots, feed himself, talk, get out of his tent, nor even out of his sleeping-bag. They regarded this misfortune as a judgement, either on him or on the whole party, for supplicating too perfunctorily the gods of the mountain.

With this sick man on our hands, with some anxiety about the safeness of the descent, and since there was now no hope of climbing the mountain and the weather was not improving, we decided on the prudent course of going down. Oliver was keen to go to Camp VI, more for the sake of treading classic ground than for any good he could do. I sympathized, and was sorry to disappoint him, for dull indeed must be the man whose imagination does not quicken at the thought of treading that ground which in its short history of sixteen years has been the scene of so much high, even tragic endeavour. A well-known passage of Dr Johnson comes to mind, though it seems almost impertinent to quote it in connexion with what is after all only a series of attempts to climb a high mountain:

> To abstract the mind from all local emotion would be impossible if it were endeavoured, and would be foolish if it were possible. Whatever withdraws us from the power of our senses, whatever makes the past, the distant, or the future predominate over the present, advances us in the dignity of thinking beings. Far from me and from my friends be such frigid philosophy as may conduct us indifferent and unmoved over any ground which has been dignified by wisdom, bravery, or virtue. That man is little to be envied whose patriotism would not gain force upon the plain of Marathon or whose piety would not grow warmer among the ruins of Iona.

CHAPTER X

LAST DAYS AND REFLECTIONS

I cannot imagine any place less suitable to choose than the high mountains, wherein to display the mastery of mankind. JULIUS KUGY

On 12 June, therefore, taking as much as we could carry, we went down by the old route. There was a traverse to be crossed on either side, but that on the western route was across ice which made the lowering of a helpless man impossible. I kept back three good men, Angtharkay, Kusang and Nukku, to help with Pasang, whose rescue, as he lay there impassive on the snow, unable to crawl, much less stand, seemed likely to prove an exacting task. The first two of these, with Pasang himself, had in 1934 comprised the happiest, friendliest, and staunchest trio with whom I have ever travelled, so it was therefore the more startling when, after some futile attempts to construct a stretcher from tent poles, they calmly suggested leaving him where he was. As they saw it, the mountain claimed a victim, and if we cheated it of Pasang then some other member of the party would be taken; for choice one of those who had taken pains to bilk the mountain of its due. My indignant splutterings were probably incoherent, but the dullest-witted must have quickly grasped the fact that the suggestion was unacceptable. Taking it in turns they carried him pick-a-back through the soft snow along the crest of the Col to the point where our fixed ropes depended. There we treated him as we did our loads, tying a bowline round the unfortunate man's chest and lowering him down to the beginning of the traverse. Crossing this was not so simple, but Nukku, who is very strong, hung on to the fixed rope with one hand, and dragged Pasang along by the feet with the other while I supported his body with another rope from above. For the rest of the way we lowered him rope's length by rope's length, dragging him where the slope was not steep enough for him to slide down by his own weight. In this fashion by midday we reached the glacier where the rest of the party were waiting. By shouting from the Col, Angtharkay had persuaded two men to come up from Camp III to help. It is to be feared poor Pasang had had a very rough passage, for his clothes were of course sopping wet, and he was half dead with cold. Now, however, he could be seated on an ice-axe which the Sherpas carried in turns by a head-strap and we soon had him in bed at Camp III.

Shipton and Smythe were waiting there, the former having now almost lost his voice since his trip up the mountain. Among the porters, Tensing too had lost his voice, but Ongdi was much better and well on the road to recovery thanks to Warren's care and the oxygen. I remember we had a heated argument that afternoon, lying about in the sun, on the ethics of the use of oxygen for other than medicinal purposes. Next day we went down to Camp I. There was some trouble getting men to take their turns carrying Pasang until the job was organized systematically and a roster devised by which each man carried him for five minutes at a time. It was a heart-breaking job that anybody would have been glad to shirk. The glacier stream was by now high, but fortunately a snow bridge was still just standing above camp. Although we reached Rongbuk on the 14th we had to stay there until the 20th waiting for the small amount of transport necessary for ourselves and our belongings. After a rest the Sherpas were employed in bringing down what they could of the stuff still lying at Camp III and West Side Camp. Most of this was left in the keeping of the monastery as it is seldom worth the cost of bringing back for the use of some problematical future expedition. Whatever is on the mountain is best left there; for it is not worth risking men's lives for the sake of tents and sleeping-bags, valuable though these are.

We discussed the question of waiting or returning in the autumn, but the discussion was largely academic as only Shipton and myself were available and as far as I was concerned there was a great deal to be done in winding up the affairs of the expedition. It might have been possible to go back to India and return later, but for such a step the approval of the Tibetan authorities would be necessary. The general opinion was that the chances of finding favourable conditions in October or November are extremely remote. The fact is not known, but assuming that winds strong and dry enough to remove the snow from the upper rocks are then blowing, there is no reason to expect any such lull as there is before the approach of the monsoon. Moreover, the wind and cold would be increasing rather than diminishing; the days becoming shorter; and, perhaps most significant, the north face receiving less and less sun. Length of daylight and warmth are both vital factors in the final climb, and Smythe's opinion is that unless there is sun on the face climbing is impossible on account of the cold.

Before leaving I wished to improve our relations with the steward of the monastery. As a result of the theft of money which took place while we were on the mountain, these were at present slightly strained. The steward had put at our disposal a room in the courtyard in which we had placed our surplus stores, including one of our cash-boxes, containing Rs. 800, enclosed inside another box. Few knew it was there, but to anyone who knew what to look for the theft presented little difficulty, so that one morning Karma Paul discovered the box broken open and the cash-box missing. He naturally complained at once to the monastery, but unfortunately gave them to understand that in his opinion the thief was one of themselves—a charge which made the steward justifiably angry. However, for a small fee, a great ceremony called 'kangso' was held in which, I gather, the thief is solemnly abjured to return the stolen goods or suffer penalties. It is a sort of comprehensive cursing or excommunication—readers of *Tristram Shandy* will remember the appalling Ernulplus' curse recited by Dr Slop while Uncle Toby whistled 'Lillibullero'. Equally terrific was this Rongbuk cursing, for a day or two later Karma Paul found the broken cash-box lying outside his tent with half the missing money. Why only half the money was returned I fail to understand, unless the thief reasoned as many people do when going through the Customs, that the declaring of only a few of the dutiable articles carried will at the same time satisfy their consciences and the curiosity of the officials. After accepting the offer of a room in which to store our things it was tactless, to say the least of it, to place the responsibility for our loss upon the monastery officials; but having assured the steward, verbally and in writing, that our interpreter's suspicions were unworthy of him and were certainly not shared by me, good feeling was restored.

The abbot invited the whole party to a meal, adding a special request that we should be merry. Although we had throughout enjoyed a sufficiency of plain wholesome food we were naturally hungry after our recent exertions and the business of eating was taken so seriously that it was not until after the seventh, or possibly eighth, bowl of macaroni stew that we remembered our host's injunction. The abbot himself dropped a hint by saying he had heard that we occasionally sang and would we mind obliging now. In fact we were not a very musical party that year and seldom gave tongue unless warmed by 'chang', but as nothing of that kind is allowed in the monastery we had to do our best in cold blood, and, which was worse, on full stomachs. Moreover, Shipton had just lost his voice and mine had not yet fully returned. By request our first number was a hymn. That was easy. I forget what it was; possibly that old favourite: 'No matter where it leads me, the downward path for me.' But then we were asked to put the mystic formula or invocation 'Om Mane Padme Hum' into lyric form. With a little adaptation we found it went very well to the tune of 'God Save the King', and we received many encores.

When our audience—a large concourse of monks and Sherpas—had suffered sufficiently for politeness, we turned to conversation which took the form of question and answer conducted through the medium of Karma Paul the interpreter. First of all we had to disabuse the minds of our hosts that expeditions to climb Mount Everest are undertaken at the instigation of and assisted by the British Government for the sake of national prestige. We assured them that this was not so and explained that Mount Everest, supreme though it was, was not the only mountain we tried to climb; that we belonged to a small but select cult who regarded a Himalayan expedition as a means of acquiring merit, beneficial to soul and body, and equivalent to entering a monastery except that the period of renunciation was short and that such admirable macaroni stew as was served in monasteries was seldom available.

Odell, who as a member of the 1924 expedition was particularly interested, then asked who had destroyed the big cairn erected at the Base Camp in memory of those who had died on the mountain that year. The abbot disclaimed all responsibility on the part of the monastery and suggested that the culprits were the 'Abominable Snowmen'. This reply staggered me, for though I had an open mind on the matter I was not prepared to hear it treated so lightheartedly in that of all places. I was shocked to think that this apparently jesting reply, accompanied as it was by a chuckle from the abbot and a loud laugh from the assembled monks, indicated a disbelief in the existence of the 'Abominable Snowman'. Such an answer in such a place, if not intended seriously, was flat blasphemy. I was soon reassured. Tibetans, like others whose beliefs are steadfast and have never been questioned, are able to treat their most sacrosanct beliefs with a frank gaiety which the outsider would deem profane. Further questioning showed clearly that no jest was intended, and we were told that at least five of these strange creatures lived up near the snout of the glacier and were often heard at night. Indeed, on one occasion some monks actually saw them. Terror-stricken, they fled to the monastery where they lay unconscious for several days—a misfortune which the sceptics among us attributed to the unwise exertion of running at high altitudes. Later we were taken over the monastery where we watched an artist painting a 'thank-ka' (temple banner) while his assistant ground a white stone into powder for paint, and finally we were conducted into the innermost shrines including that reserved for the devotions of the abbot himself. In one of these was a large lump of greenish black rock, probably crystalline, measuring about 18 in. cube and weighing perhaps ½ cwt. On its smooth flat surface was the very clear impress of a large human foot. It was said to have been found in the vicinity of Camp I and was evidently regarded with some awe. The imprint appeared to be genuine enough and the fact that the stone was kept in one of the innermost shrines and not exposed to the admiration of the vulgar seemed to rule out the likelihood of its having been carved by hand.

Now Odell, who was with me, is a geologist of some standing, so with my childlike faith in the omniscience of scientists I turned confidently to him for the answer to the enigma. How, when, and where, did the stone receive this remarkable impress, and who made it? The Oracle was dumb and spake not. Not so much as a few technical words about the stone itself, about igneous rock intrusive in calc-gneisses, cretaceous eocene limestones, or compressed synclines caught up in folded Jurassic schists, and the rest of the jargon with which the geological pundit baffles and dumbfounds the humble inquirer. Not even the usual tantalizing ambiguous couplet beloved of Oracles which I myself could have supplied at a pinch had I been asked, as for example:

> Whoever trod upon this stone,
> Was a thing of flesh and bone.
> If you will any further know,
> The secret's hidden in the snow.

Having been found in the moraine near Camp I the stone must have been brought down by the glacier and deposited there at some very distant date. We know for a fact that no Tibetans ever traverse the glaciers, least of all the Rongbuk which leads nowhere and which owing to its proximity to Mount Everest they regard with special awe, and we are driven to associating this footprint in some way with people whose habitat is above the snow line, in other words the 'Abominable Snowmen'. But I will not press the argument. A scientific friend (for I am not without them) points out that the footprint was probably fossilized and may have been made at a recent date in the geological time scale—the middle tertiary—when the northern part of the Himalaya and most of Tibet were covered by a great sea, called by geologists the 'Tethys Sea', in which deposition of sediment had continued for ages. And there the matter must rest until some future visitor to Rongbuk shall tell us whether the stone in question is crystalline or sedimentary. In the Himalaya any strange footprint inevitably gives rise to a certain train of thought, and this print is one of the strangest. Since the subject is one of interest to scientists, and since no book on Mount Everest is complete without appendices, I have collected all the available evidence, old and new, and relegated it to the decent obscurity of Appendix B.

As his leave was up Oliver left ahead of us on the 17th. On the 20th the rest of us began the march back—impatient of delay, disappointed at what had been accomplished, but satisfied that the expedition had not been without its lessons. Of these I think the most important was to show that a small expedition, of seven or even fewer, costing less than £2,500, has as good a chance of climbing the mountain as one of a dozen members costing £10,000 or more. Expense, of course, is not an absolute criterion. But we live in a commercial age and as a measure of relative efficiency, when the accomplishment has been very nearly equal, it is a fair yardstick

21. Khartaphu (23,640 ft.), from the Kharta glacier (p. 61)

22. The western face of the North Col—showing debris of great avalanche (p. 79)

This modest claim that such an expedition stands at least an equal chance will be substantiated by members of the party who agreed that the methods employed were sound; but I like to think that enough was done to convince all who are interested in these matters. At any rate mountaineers will not dispute that such methods, if they can be employed, are preferable and more in accord with mountaineering tradition and practice. In spite of very unfavourable conditions two parties of two—a third could have been sent up had conditions warranted—were put in position to make a bid for the summit. Nor were these men weak or worn out through overwork and rough living. No hardships had to be endured through inadequate equipment or insufficient food. The coughs, colds, sore throats and influenza which troubled us also troubled those expeditions equipped and fed in a comparatively luxurious way regardless of expense. As Mr Shipton says of the attempt by Mr Smythe and himself: 'It was bitterly disappointing, for we were both far fitter at these altitudes than we had been in 1933, and the glittering summit looked tauntingly near.' And again, in a discussion at the Royal Geographical Society[1] subsequent to the expedition he said: 'As regards the health of the expedition, I am convinced that the party kept far fitter than in 1933, when two members were confined to their beds with 'flu for a fortnight, one other was sent to Kharta with bronchitis, another was incapacitated for months with a gastric ulcer, and we all suffered more or less severely from laryngitis, colds, etc. No party has ever been at full strength on Everest.'

Of course the organization was not perfect. Some thought that the process of rigorous economy had been carried too far, others that it had not gone far enough. Unanimity in the matter of food, for example, is no more to be expected among seven members of an expedition than it is among seven members of a family. Some grumbles were heard, but more can be heard in the dining saloon in any passenger ship where the choice of food is as bewildering to the mind as its assimilation is to the stomach. Those who have experienced or read about the large expedition (whose members are often free of any financial concern in it) have a different scale of values from those who have not. The necessity of one man

[1] Vide Appendix A.

becomes the luxury of another, but with a little give and take a suitable compromise can generally be reached. For instance, if a man expects to have a choice of three or four kinds of marmalade for breakfast it is a disagreeable surprise to find none at all. The important thing is to keep firmly in mind the essentials, and then to make a few, very few, concessions to human frailty; not, as is often done, to include in the expedition stores everything which might conceivably at some time be wanted by somebody—a process which is often undeservedly called thorough organization. The French Revolutionary armies, destitute of most things needful and with twisted hay-ropes for boots, might justifiably think or even hint that their equipment was in some respects deficient; but the Convention Representative with the army (the Political Commissar of those days) was deaf to complaints of that sort and met them with the laconic remark: 'With steel and bread one may get to China.' I do not mean to suggest that with pemmican and an ice-axe one can get to the top of Everest, but a touch of this spirit is not amiss in the organizer of any expedition and will form a wholesome corrective to any tendency to over-equipping and over-feeding. Mr R. L. G. Irving maintains this view: 'And do we destroy nothing by using all this mass of men and material to conquer Everest?' he asks; Mallory, after a catalogue extending to some forty lines of the various items in the vast collection of stores and equipment carried across Tibet, concludes: 'When I call to mind the whole begoggled crowd moving with slow determination over the snow and up the mountain slopes, and with such remarkable persistence bearing up the formidable loads; when after the lapse of months I envisage the whole prodigious evidence of this vast intention, how can I help rejoicing in the yet undimmed splendour, the undiminished glory, the unconquered supremacy of Mount Everest?' And what mountaineer would not agree with Mallory if the mountain is only to be won by the skilful use of material carried to the *n*th power? Mr Irving adds: 'Everest will be conquered by just the very thing in which the present age excels, the skill to use the material things that nature has provided.' Let us hope he will be proved wrong; the 1938 expedition will have served a useful purpose if it has done nothing more than show that this need not be so.

Apart from unnecessary elaboration of food and equipment there are, amongst the material resources now used in Himalayan mountaineering, more baleful things, like aeroplanes, wireless, and oxygen apparatus. For some years past aeroplanes have been used in Alaska for carrying men and stores to a base, and enabling a reconnaissance to be made in a few hours instead of a whole season, the time required by older methods of transport. There, the great distances, the shortness of the season, and the absence of any porters, are at least reasons if not excuses for using a method which takes account solely of the end to be attained—the rapid subjugation of some peak. It is akin to the use of cars and lorries by shooting safaris in Africa; at first they were actually used for shooting from because it was so much easier to approach the game; but as public opinion was sufficiently strong to stop this they are now merely used to convey the party to its ground, and to make sure that the members derive neither enjoyment nor benefit from their short excursion into the wilds. In 1938 an aeroplane was used in the Himalaya to drop loads near camps on Nanga Parbat. The success may not have been great but the technique will soon be improved; indeed, I can safely say has been improved, for during the war the dropping of supplies to inaccessible places, even by night or in mist, was brought to a high state of perfection. Doubtless some of our most progressive thinkers are already toying with the idea of dropping men as well somewhere near the summit. As a dropping ground the north face of Everest leaves something to be desired, but I can imagine it being done on a mountain like Nanga Parbat with some degree of safety. Still, it will probably be found safer and quicker to climb there in the end, and if we must move with the times then we must give up mountaineering, or at least cease calling what we do mountaineering, for when such adventitious aids are used that it certainly is not. If our end is just to plant a man on top of the mountain then I suppose any means are justified, but if our end is mountaineering in the true sense then we should stick to the rules. All's fair in war and the habit of talking of the assault and conquest of a peak may lead us to think that the same holds good for mountaineering and that mountains are foes to be subdued rather than friends to be won. There is a good case for dropping bombs on civilians because so very few of them can be described as inoffensive, but mountains can claim the rights of 'open towns' and our self-respect should restrain us from dropping on them tents, tins, or possibly men.

Then with regard to wireless it may be thought that a foolish prejudice against it by one holding old-fashioned views robbed the party of a possible chance through their ignorance of weather changes. Such was not the case; nor, I think, will it be for some time. Mountains, particularly the big Himalayan peaks, make their own weather, and at present the wireless forecasts are based on weather conditions prevailing hundreds of miles away from the Himalaya. Dr Sen, of the Indian Meteorological Department, thinks that even with the meagre information at present available it should be possible by February of any year to foresee any exceptional acceleration or retardation of the advance of the monsoon. Such information would of course be useful, but Everest is a special case. It is not possible to wait until February to decide whether an expedition will take place or not. If permission is received for a particular year the expedition must go that year, and if it knows that the monsoon will probably be early then it merely knows that its chances are greatly diminished, for the weather will probably be too cold to do anything beforehand. But this is rather different from the expectation of basing climbing plans upon weather forecasts when actually on the mountain. A German party on Nanga Parbat in the same year had wireless and were defeated by the weather at 23,800 ft.; an American party without wireless, climbing on the difficult and less well-known peak of K 2 only 120 miles away, reached a height of 26,000 ft. in perfect weather.

Knowledge of the onset and progress of the monsoon is of interest but it is not enough. The phenomena known as 'western disturbances' seem to be what usually upset the apple cart. If a succession of these are encountered from early May onwards then the attempt can be written off, be the monsoon early or late. On 30 April 1936 and on 5 May 1938 one such storm turned the mountain white and in neither year was it ever again in climbable condition. One has only to read Mr Ruttledge's account of what happened in 1936 to realize how useless are weather forecasts to the leader. On 30 April, three weeks before the schedule date for establishing the North Col Camp, they received wireless warning of the approach of a 'western disturbance', and with remarkable punctuality it began to snow at 3 p.m. the same afternoon. Naturally, this premature snowfall did not cause them to alter their plan; it has been the opinion of every expedition that in early May the cold is too severe for high climbing, and in any case the damage was already done. Justifiably enough, from all previous experience, they expected this early snow to be blown off by westerly gales before the onset of the monsoon itself, and this expectation was disappointed, as ours was.

To take our own case; supposing in the last days of April we had been warned of the approach of bad weather I do not think we would have altered our plans. The cold at Camp III was a convincing deterrent to going any higher and we would have gambled on the very reasonable chance of there being enough wind to clear the mountain before the arrival of the monsoon at the beginning of June. The leader would have been given the invidious choice of having some of his party disabled, perhaps permanently, through frostbite in a desperate attempt to forestall the weather, or of ignoring the forecast and being subsequently reviled for his unheeding contempt of science. Even if some omniscient being at the other end had informed us by wireless that 'western disturbances' would follow in rapid succession until the beginning of the monsoon proper (which was what happened), we should certainly have been well informed but quite unable to do anything about it except perhaps pack up. The long-range forecasts of the probable date of the monsoon appearing, which Dr Sen considers possible, are not really of such importance as a forecast of the weather in the pre-monsoon period, the month of May. And I doubt if this is possible even if we submit to the stipulated establishing of a meteorological station with its transmitting set in the vicinity of the mountain. With the mass of weather data available in Europe can anyone tell us, for example, what sort of a season we shall have in the Alps? Will July or August be the best month? How much easier it would be to arrange our climbing holidays if they could. No, mountaineering, like farming, sailing, and all our most interesting activities, is a chancy business ruled largely by the weather which we take as it comes as philosophically as we can; unless or until we make an idol of 'success' and stake too much upon it in money or material to submit to the ordinary rules, when we wish to become like Hamlet's politician 'one that would circumvent God'.

The recent war has introduced us to wireless receiving sets which can be slung from the shoulder like a water-bottle and which weigh very little more. Although by doing so the party lose one outstanding advantage of being in Tibet, should they wish to have their thoughts and feelings harrowed by listening daily to the news I see no reason why one of these should not be taken, particularly as the volume is such that no one need be compelled to listen. If such sets were to be used for receiving weather forecasts from Calcutta special arrangements would have to be made with the Indian Meteorological Department and an Indian broadcasting station, and if such arrangements could be made for the reception of forecasts on one of these pocket sets weighing 5 or 6 lb., well and good. The forecasts would be of interest if not of use, and would at any rate excite great degrees of what Hazlitt called the great springs of life —Hope and Fear—and so prevent the emotional faculties of the party from becoming dulled. But whatever value there may or may not be in weather forecasts I believe there is no justification for the taking of a big set with batteries and engine such as was used in 1933 and 1936. It is costly, difficult to carry, and involves the addition of one or possibly two mouths to feed; while if it is to be used for sending messages on the progress of the expedition, the leader, in order to justify its presence, has to betake himself ardently to journalism.

Very small sets for short-range talking as well as receiving are also available now. Their weight, though small, is appreciable, and above the North Col every pound counts. It is easy to imagine circumstances in which they would be useful, but none in which they would be indispensable. In my opinion they would not be worth their weight.

An expedition that sells itself to a newspaper is liable to find itself saddled with a wireless set as a result. Nearly every one admires and upholds a free, uncontrolled and uncontrollable Press, throwing the searching beam of publicity into obscure corners and generally taking care that the State receives no hurt; but when the fierce beam lights upon the mountains, the people climbing them, or better still falling off them, then our admiration is apt to be withheld. The Press, with some honourable exceptions, has no soul to be saved or body to be kicked, and neither violent protest, savage scorn, nor corrosive sarcasm, will turn it or its minions from what is mistakenly conceived to be the path of duty. All this was acknowledged and regretted in more temperate but sufficiently plain language by one of the few responsible Dailies left. A leading article declared that: 'The standard of intelligence, taste, and accuracy of the Press to-day is admittedly not what it was 30 or 40 years ago. This may be merely a reflection of lower standards on the part of the public which the Press itself has done its best to lower. Either way the responsibility cannot be escaped.' The evil is there and we can do nothing about it; only time, the great healer, will diminish or worsen it.

Whether an expedition makes an agreement for the sole rights of the story with one particular newspaper or not, every newspaper will join in the hunt. If no agreement at all were made, the position would be worse, for there would be no untainted source to which the serious inquirer could turn for reliable information. Possibly if the contract were made with some news agency, which would supply news to all newspapers there would be no obligation upon any paper to regale its readers with sensational and distorted accounts. Such an arrangement might even mitigate the persecution of relatives of members, another unpleasant aspect of publicity at present as common as it is deplorable. At any hour of the day or night relatives may be rung up and asked to provide news or comment on its absence. The slamming down of the receiver before a word has been said will not prevent the appearance next morning of some fatuous remarks wrongly attributed to the victim, whose cup of bitterness is still to be filled by a tart reminder from the man on the spot that nothing must be said to the Press.

Something more was learnt about the use of oxygen and the apparatus itself. Opinion amongst the party as to the ethics of using oxygen was about evenly divided; few thought that in its present stage of development it was of much value. My own opinion is that the mountain could and should be climbed without, and I think there is a cogent reason for not climbing it at all rather than climb it with the help of oxygen. It is only fair to say that many men of far greater authority to speak for mountaineering tradition and the ethics of climbing take an opposite view to mine. The late J. P. Farrar, for example, who in 1917 to 1919 was President of the Alpine Club, and from 1920 to 1926 Editor of the *Alpine Journal*, stated in a review of the 1922 expedition that 'this objective, the conquest of the mountain, must be kept steadily in view, and its attainment be attempted...with *every available resource*'. This is the conclusion of an article largely devoted to the oxygen problem in which he answered the question 'Would it be unsportsmanlike?' by saying: 'Yet Everest is to be allowed to clothe itself with air containing a far less proportion of oxygen than is needed for the development of the full powers of man, and the mountaineer who attempts to make good the deficit is held to create conditions so artificial that they can never become legitimate mountaineering.' This seems to me equivalent to saying that since Everest allows itself to add several thousand feet to its stature beyond that of any ordinary mountain we can make good the deficiency of our climbing powers by using a man-lifting kite; or since some particular rock face assumes an unwarranted degree of steepness we will therefore drive in a few pegs. The views of a representative body of mountaineers have never been taken, but, ethics apart, my feeling and my principal reason for opposing it, is that a successful oxygen climb would only inspire a determined wish to repeat the climb unaided. There would be no finality about it and we might see, always assuming the consent of the Tibetan authorities, another long-drawn-out series of attempts such as we have already had. Man's competitive instincts are fortunately not easily tamed; until the South Pole was reached rivalry was intense, but once it was reached honour was satisfied. Mountaineers are never content until a mountain has been climbed by every possible route, but since there are probably no alternative routes on Everest I feel sure that if it were climbed with the help of oxygen there would be for mountaineers an instinctive urge to climb it again without. The plausible argument that it is better to climb the mountain with oxygen rather than not at all is therefore unsound. And it is rank materialism unless we assume charitably that those who think thus believe that the mountain will never be climbed without oxygen, and hope, as we do, that once it is climbed the striving for height records, the publicity, the national rivalry which was a feature of

23a. The Traverse and final climb to North Col (p. 71)

23b. The climb to Camp V (p. 81)

24. Cho Uyo and West Rongbuk glacier from Camp V.
(Panoramic with Frontispiece) (p. 82)

the thirties, will then cease and that the Himalaya will become the playground they should be. Mountaineering is analogous to sailing, and there is not much merit to be acquired by sailing with the help of an auxiliary engine. If man wishes gratuitously to fight nature, not for existence or the means of existence but for fun, or at the worst self-aggrandisement, it should be done with natural weapons. Obscure though some of them may be, the reasons which urge men to climb mountains are good enough reasons for wishing to climb the highest mountain of all, provided it is done in the normal way of mountaineering by a private party responsible only to themselves. The various other reasons which have been adduced in the past, such as demonstrating man's 'unconquerable spirit', or increasing our knowledge of man's capacity, should not persuade us to alter our methods, especially when by so doing we stultify the reasons themselves. I take it that when a man has to start inhaling oxygen his spirit has already been conquered by the mountain and the limit of his capacity has been very clearly defined. Even the hard-worked but astonishingly powerful reason, the advancement of science, has been brought to bear. As the late Sir Francis Younghusband wrote in an introduction to *Mount Everest: The Reconnaissance*: 'No scientific man, no physiologist or physician, can now say for certain whether or not a human body can reach a height of 29,000 ft. We know that in an aeroplane he can be carried up to a much greater height. But we do not know whether he can climb on his own feet to such an altitude. That knowledge of men's capacity can only be acquired by practical experiment in the field.' I do not know whether it is so, but if the scientists are interested in their cold-blooded way in a man's behaviour at 29,000 ft.—whether he dies a quick or a lingering death, for example—then another unanswerable objection to the use of oxygen presents itself, for the means defeats the purpose. Far be it from me, or from any mountaineer, to balk our scientists. But they want to have it both ways. The physiologist may be genuinely interested in a man's capacity at great heights, but other of his fellow scientists are more interested in equipping a man with some device that will overcome natural difficulties. And, as modern developments show, it is an engaging but fatal characteristic of scientists that once presented with a material problem of this kind they set about the solution with a single-minded devotion that excludes any other considerations whatsoever, be they ethical, humane, or merely of common sense. This fanatical approach to a problem is pleasantly illustrated in the following statement made in 1922 by one who is a mountaineer as well as a scientist:[1] 'If this year's expedition is not successful, some enthusiastic millionaire may yet provide a liquid oxygen generating plant near the mountain which will reduce the ascent of Everest to a question of *£. s. d.*' Thus the frenzy of the scientist readily extinguishes the common sense of the mountaineer and raises a very ugly head indeed. As the Spanish proverb says: 'Science is madness if good sense does not cure it.' The point about liquid oxygen is, I believe, that it would be much more portable, but owing to its rapid evaporation it would have to be manufactured on the spot.

In reply to any ethical objections that one may raise the oxygen enthusiast merely points significantly to one's clothes, well-nailed boots, snow-glasses, or ice-axe, all of which he considers sufficiently artificial to condone the use of yet another artificial aid, namely oxygen. And it is a difficult argument to refute. An appeal to common sense merely invites the retort that it is precisely that to which he is making his appeal. 'Why', he asks, 'jib at using oxygen? If your oxygen could be provided in the form of pills you would use it quick enough, just as some of you use pills to make you sleep.' Well, perhaps one would. Oxygen pills, one hopes, would not weigh 25 lb. But it will be time to decide that knotty point when the pills are forthcoming; meanwhile, if we are being illogical, which I doubt, let us continue being so; for we are an illogical people and mountaineering is an illogical form of amusement which most of us are content to have as it is. It is a case of the pot calling the kettle black. The gas school contend that it is very doubtful if the mountain can be climbed without oxygen. But they hasten to assure us that even with the aid of oxygen it will be sufficiently difficult; that all the difficulties, physical and mental, will not be thereby abolished, or even greatly diminished. Possibly not; in the opinion of some they will be increased, but here they are themselves illogical; for the point is that with a little more scientific research and applied ingenuity they hope and expect to produce the perfect apparatus which will go nearly all the way

[1] Mr P. J. H. Unna.

towards abolishing fatigue, breathlessness, cold, the benumbing mental effects of high altitudes:[1] in fact the very foes with which the mountaineer takes pleasure in grappling, and those without which Everest, in particular, would not be the very redoubtable mountain it is. Possibly those who take a different view on the use of oxygen and aeroplanes as accessories to Himalayan mountaineering will say that on these matters instead of doing a little unaccustomed thinking I have been content to consult merely my feelings. Even if this were so, and I admit that thinking is an unusual and difficult exercise for me, my views are not on that account worthless. We have not to go far from England or far back in history to see the effects of using reason alone and neglecting to consult human feelings. A far greater authority, the distinguished author, editor and educationalist, Mr George Sampson, has laid down: 'Reason looks well on paper: but in reality we have scanty grounds for assuming that reason is a better guide to life than feeling.'

But reverting from theory to practice I can fairly say that on this year's showing the advantages conferred by using oxygen did not outweigh the disadvantages attending its use—difficulties related to the weight itself, to the unbalancing effect of that weight on steep ground, and the number of extra porters required to carry up cylinders. The unbalancing effect will become really serious on the crucial part of the climb—Norton's Traverse, the Couloir and the final pyramid.

Oxygen apparatus, wireless sets, or aeroplanes, are not the sort of things one expects or delights to find with a mountaineering party, of which the keynote should be simplicity. No lover of mountains would care to come upon such things (a liquid oxygen plant for example) lurking obscenely at the foot of the 'Delectable Mountains' of his dreams; but since the Himalayan giants, Nanga Parbat, Kangchenjunga, K 2 and Everest, have repulsed all attempts upon them man seems more and more inclined to resort to scientific aids in order to force success and to reassert his superiority over

[1] As long ago as 1922, before anyone had shown what could be done without oxygen, Professor Dreyer, who, with G. I. Finch and P. J. H. Unna, took a principal part in developing the idea of using oxygen, confidently expected 'that the artificial supply of oxygen will make the climbers as physically fit as they would be at some altitude between sea level and 15,000 ft.'

nature. Since fair means are ineffective we begin to think out foul means, and treat a mountaineering problem as a scientific problem, or what the Americans would call a problem in logistics—how to get a man by any means to the top of Everest.

As every expedition has learnt, the most important factor in climbing Everest is the weather and conditions near the summit. Without favourable conditions, no matter what auxiliary aids are employed, the mountain will never be climbed, and with them it will probably be climbed unaided. In 1938 we found that even after the monsoon had deposited a generous layer of snow on the mountain it was possible to reach a height of twenty-seven thousand odd feet, but no higher. This was merely a reaffirmation of what had been categorically stated by Norton in 1924 and by Smythe in 1933—that snow on the upper rocks is fatal to success. The climbing of many mountains after snow has fallen (when a few days have elapsed) is easier, but near the summit of Everest snow behaves in a different way and appears never to consolidate. That it does not is obvious, for otherwise the upper rocks would not be entirely devoid of snow in the early spring as they always are. In 1938 there appeared from below to be less snow on the upper rocks than there actually was; but though I think we realized that what snow there was was quite enough to damn us, some of us went up with the lurking belief, or at any rate hope, that we should find the snow hard; that this incontrovertible theory was wrong, that the snow behaved in the normal way and that it was merely the wind which was abnormal, blowing in the winter with sufficiently searching ferocity to tear off a layer of several feet of hard frozen snow; but alas:

> The heart of man has long been sore and long is like to be,
> That two and two will still make four and neither five nor three.

We found the snow as loose and powdery as on the day it fell, with no sign at all of consolidation or of adhesion to the rock beneath, thus forming a fatal barrier. In such conditions the enormous effort involved in moving at all, the correspondingly slow rate of progress, and the danger from avalanching snow, are insurmountable difficulties. At that height, prolonged siege tactics, methodically clearing away the snow, are not to be thought of; and if they were, the avalanche danger still remains. If in order to be sure of finding the summit clear of snow the party makes its attempt too early it is met with the equally fatal and more cruel impediments of wind and cold, which, of course, become more severe as the climber ascends, who in turn becomes less able to withstand their effects owing to oxygen lack.

The approach by the west side of the North Col has already been discussed. Perhaps if the old route had behaved as we expected, and as it often does, treating us to an exhibition of 'frightfulness', we might have been more enthusiastic about the alternative route. Before monsoon conditions are established the old route is safe and nothing is gained by using the slightly longer approach; even after the onset of the monsoon, except in conditions considered certain to result in the formation of wind-slab on the east side, I should hesitate about using the other. The only really large avalanche we saw that year fell from the west side and there were spells of strong westerly winds blowing snow over the North Col and depositing it on the east side which should have resulted in the formation of wind-slab but which apparently did not. Uniform regular slopes such as on the west side are, I should think, more prone to avalanche than broken, crevassed slopes. On the other hand the east side is steeper. But in the present state of our knowledge Himalayan avalanches are as unpredictable as they are vast, and though one's judgement will probably be faulty it is at least satisfactory to have two alternative lines of advance and retreat upon which to exercise it.

In conclusion I should like to record my gratitude to the Mount Everest Committee who placed their confidence in us; to the friends known and unknown who generously subscribed; and to all members of the party, European and Sherpa, who did so much. Of the future little can be said. No one who has been high on the mountain, though very aware of the appalling difficulties, affirms that it is impossible to climb it given the right conditions—no snow on the upper rocks and a sufficiently warm, windless day. Such conditions occurred in 1924 and 1933, and will no doubt occur again. The job is taking longer than anyone expected, but it is not impossible, or at least it is in accord with Nansen's definition of that word. That great man and great explorer defined the 'difficult' as that which can be done at once, and the 'impossible' as that which may take a little longer. The climbing of Everest is evidently one of his 'impossibles' which is taking a little longer.

No doubt more attempts will be made but after the bitter disappointments of 1936 and 1938 it is clear that the odds against meeting perfect or even favourable conditions on any one particular visit are longer than was thought. If future expeditions were to be carried out on the lines indicated in 1938, there should be no great difficulty in financing more frequent expeditions. The ideal to aim at would be a consecutive series of three, four, or five attempts. But we should make up our minds whether they are to be mountaineering or scientific expeditions, not a combination of the two, both for the sake of not diminishing the sufficiently scanty chances of success and for maintaining the true tradition of mountaineering. If the scientists really wish to carry out high-altitude tests in the field they could not make a worse choice of laboratory than Everest. For political reasons it is extremely inaccessible and the opportunity granted by the weather for climbing high upon it is fleeting. Moreover, there are two other mountains less than a thousand feet lower to which they could go every year were they so minded with no one to say them nay.

I am not asking that access to mountains should be denied to scientists, or that anyone found upon a mountain making scientific observations should be forthwith abolished more or less painlessly according to the purity of his motives. In that case we should soon have to regret the demise of many of our most ardent mountaineers, the Alpine Club would be decimated, and to the long list of martyrs in the cause of science we should have to add the familiar names of de Saussure, Tyndall, even Whymper himself, whose adventures with a mercury barometer in the High Andes are the comedy after the tragedy of the Matterhorn. No, I merely ask that mountaineering and science should be kept distinct, in particular that the problem of climbing Mount Everest, like any other mountain, should be left to mountaineers to solve, and that those actively engaged in solving it should not be expected to enter what Goethe calls the charnel-house of science. For it is only on the biggest mountains that we have to be on our guard against the encroachment of science. Without the aid of any formal rules, climbing at home and in the Alps is much as it was (except of course in standards) when first begun, and the earliest traditions of the Alpine Club are common to all and upheld by all; and this because the droves of mountaineers with a scientific bent, and the mountaineering scientists, while enjoying themselves in company with their less gifted brethren, and at the same time wielding geological hammers, swinging thermometers, or boiling them, have studiously refrained from suggesting any mechanical or scientific aids to lessen the ardours of their studious pleasure. Admittedly some provocation was lacking—there was no mountain high enough to tax their lung power—but it goes to show that a scientist with a feeling for mountains is not so utterly forsaken as one without, that his humanity is not yet extinguished, and that a powerful tradition may still be respected. May it ever be so, and may the provocation offered by the problem of Mount Everest be firmly withstood.

Whatever we may propose with regard to Mount Everest it is the Tibetans who dispose. We are under an increasing debt to the authorities at Lhasa for their goodwill; for permitting men whose motives, if not suspect, are at any rate incomprehensible, to visit their wonderful mountain—not only permit them to visit it but afford them powerful help and kindly hospitality. Their friendliness in the past encourages us still to hope. Meantime let us count our blessings—I mean those thousands of peaks, climbed and unclimbed, of every size, shape and order of difficulty, where each of us may find our own unattainable Mount Everest. And may those of us who have tried and failed be forgiven if we ask ourselves the question put by Stevenson: 'Is there anything in life so disenchanting as attainment?'

APPENDIX A[1]

Mr E. E. Shipton: I am very pleased to have this opportunity of congratulating Tilman on the way in which he ran the expedition, not at all an easy matter from various points of view. The great bugbear of an Everest expedition is that, whereas on an ordinary scientific exploratory expedition parties can be sent off into an unexplored area, each member having his own particular job, on Everest at least 75 % of the time is spent in doing nothing. That fact is not generally realized. Members tend thus to become bored and to criticize unimportant things. It is difficult to organize the party so as to avoid this. The expedition this year was a pleasant one; we were all socially harmonious, and I think we all worked together well. But the question of doing nothing for long spells of time is a difficult one to get over. If I tell you that bed-sores, both physical and mental, are a greater hardship than altitude you will probably regard it as a joke; but there is quite a lot of truth in the statement.

As regards the health of the expedition, I am convinced that the party kept far fitter than in 1933, when two members were confined to their beds with 'flu for a fortnight, one other was sent to Kharta with bronchitis, another was incapacitated for months with a gastric ulcer, and we all suffered more or less severely from laryngitis, colds, etc. No party has ever been at full strength on Everest.

Although on the expedition in 1935 we tried out, with success, the small expedition idea, there are still a number who regard it with grave suspicion, but the 1938 expedition has done much to prove the contentions of those who advocate the small expedition. I now feel that if I were to argue further in favour of the small party, I should be flogging a dead horse. I should however like to draw attention to the excellent American expedition to K 2. They cut their food and equipment down a great deal further than we did and I think they gained a greater measure of success on the second highest mountain in the world; a mountain which had looked so difficult that the Duke of the Abruzzi had, thirty years previously, given it up without making a serious attempt to climb it. Until the American party went there in 1938 no other serious attempt had been made. That expedition reached an altitude on an almost completely unknown and extremely formidable mountain of about 26,000 ft.: a very remarkable feat.

I do not think there can be any doubt that Mount Everest will be climbed one day, but there must be perfect weather and perfect conditions otherwise, a combination which is not nearly so common as earlier expeditions appear to have thought.

I think Tilman has told you everything there is to be told, and beyond congratulating him again on the splendid way in which he ran the expedi-

[1] Discussion reprinted from *Geographical Journal*, vol. XCII, No. 6, Dec. 1938, pp. 490–8.

25. Looking eastward from 27,200 ft., near Camp VI: Gyangkar Range in far distance (p. 83)

26. Looking east from Camp V (p. 82)

tion and the excellent manner in which he has presented it to you, there is nothing more for me to say.

Mr F. S. SMYTHE: I should like to associate myself with everything Tilman has said with regard both to the small expedition and to the oxygen. There is a point which has impressed me very much. As you know, oxygen is supposed to be, and is, the best possible treatment for frost-bite, and yet apparently Lloyd, who was carrying oxygen, suffered from the cold just as much as did Tilman.

Another interesting point which perhaps has not been stressed is that of the cold at high altitudes. A difficulty both parties experienced was the cold in the shadow on the north face of the mountain, and I will go so far as to say that I am quite sure as a result of my own experience that the mountain cannot be climbed in shadow. I remember watching the upper part of the North Face from Camp 2 in April, and as far as my memory goes the sun did not get on to it until about eight to nine in the morning. That is too late to start from Camp 6 for an attempt on the summit. Any suggestion therefore as to a post-monsoon attempt must be defeated by that one fact alone, and therefore the possibility of climbing Everest is definitely limited to that period, which we did not get, just before the monsoon.

Then as to the question of the food: Tilman alluded to the impossibility of eating anything above Camp 4, but in the next breath he said that he and Lloyd ate pemmican at Camp 6, and to any one who has eaten pemmican, I would almost say at any altitude, that is a remarkable circumstance. The usual procedure is to plan out some food for high-altitude camps, and when one gets to Camp 3 one lumps as much as one can remember of it into a rucksack and takes it up. The question of high-altitude food should be a matter for experts and for all the members of the expedition. It should be documented at the start and taken up in small boxes, each box being labelled for its particular camp. In that way the climbers might conceivably get something they had not eaten before, and that would help to get them to the top.

Dr C. B. M. WARREN: You may have gathered from what Tilman has told you that he has a supreme distrust of scientists and doctors. I never quite knew whether to be flattered at being asked to go on the Everest expedition because I spent my spare time mountaineering, or whether to be insulted because Tilman asked me to go as a doctor. In spite, however, of Tilman's views about doctors in general, I shall dare to talk to you about the medical aspects of the expedition.

To begin with, we had the usual crop of sore throats, and I think Tilman is quite right when he says there is nothing very much that we can do about these once they have developed. We tried throat sprays and painting the throat, and I tried wearing the Matthews respirator, with which I think I had a certain amount of success. But the suggestion that I put forward, perhaps for future occasions, is that the leader only or a competent transport officer should lay the camps up the East Rongbuk glacier as far as Camp 3 when the weather is still very cold, and the rest of the party should be kept down in the Kharta valley at reasonable levels where they will keep fit and can practise mountaineering. Thus they would not be upset by the high altitude and the dry atmosphere until they were ready to make the actual attempt on the mountain.

In spite of the fact that a doctor on the expedition has been decried as unnecessary, we had at least one very serious medical casualty this year in our porter Pasang, who after going up to Camp 6 became completely paralysed down one side; he could not walk at all, and had to be carried and lowered the whole way down the slope of the North Col. This was most unexpected in a young man, and the question was whether the paralysis had anything to do with the altitude or whether it was just fortuitous. I cannot help feeling that there was probably some connexion between the two. Certain accidents which occurred on past expeditions, when reviewed in the light of Pasang's illness, might perhaps have had a similar cause. When people go to high altitudes alterations occur in the blood as a result of acclimatization. It is not at all certain that these alterations are entirely beneficial. It may be that after a certain time there is clotting in the vessels of the brain because the blood has become more viscous, and this would cause serious accidents such as this case of paralysis.

I would like to support Smythe's remarks as to the diet. I think it is most important on Everest, from the medical point of view at any rate and because we want to keep people from getting what we call intercurrent infections, that they should be well nourished. Not only should they have their vitamins, which is easily managed nowadays, but they should have an adequate diet. As it is almost impossible for members of the expedition to eat a proper diet at high altitudes above Camp 3 I think it is all the more important that they should have a very carefully worked out diet at the lower levels. When we went over to the Kharta valley this year we had to travel light because we were going over a high pass; but the idea of going there was to recuperate. And yet we seemed to spend most of our time feeling hungry on the way down there and feeling hungry when we had reached the valley of plenty. I suggest that in future a dietician should always be consulted before an expedition goes out. After all, he has spent most of his life carefully studying diets, and even though you do not accept what he says you would at any rate get expert advice.

I have been asked to say a word or two about oxygen, but Lloyd will give you more indication as to whether it was successful on the mountain or not. We had two kinds of oxygen apparatus with us. In one type the wearer breathed pure oxygen and all the external air was excluded; in the other type he breathed oxygen which was diluted with air. Both types of apparatus had been tried out at home, but I made special tests with the newer closed type, in which only oxygen is breathed. In 1937 I took it out to the Alps and climbed the Matterhorn in it, the Wellenkuppe, and one or two other mountains, to see whether one could really climb steep rocks. Mechanically the apparatus worked extremely well. The only drawback was that those mountains were not high enough to enable one to decide whether there were any beneficial effects from the use of the gas. So we took out both types of apparatus this year. When I went on the North Col the first time I tried the closed apparatus, and what Tilman has said about my reactions is more or less accurate. When wearing the apparatus I was completely free of fatigue in going uphill, but there was a feeling of suffocation of which it is hard to localize the cause, unless it lay in the fact that the design had been altered slightly at the last moment. I do not think that one can entirely rule out of court the closed type of apparatus on that account; I think it should be given further trials. The other type of apparatus I was not able to try myself because I had to go down off the Col, but Lloyd will tell you how he climbed above Camp 6 in it, and give you his reactions.

Mr Peter Lloyd: We have heard a lot of hard words this evening about oxygen, not only from Tilman but also from Smythe. I have a lot of sympathy with the sentimental objection to its use, and would far rather see the mountain climbed without it than with; but on the other hand I would rather see the mountain climbed with it than not at all. Climbing Everest is not, after all, like a game of cricket, for which you can make rules as to what is fair and what is not. It is a big enough proposition to demand all the forces that we can bring to bear against it.

Warren has told you of, and Tilman also mentioned, the two types of apparatus that we had with us this year. It was a distinct disadvantage in many ways having to take both types with us, but having them both there it was obviously desirable that we should try to test them under the actual conditions above the North Col. That is the only test that is of any value.

It is difficult to make adequate tests under those conditions, and I am fully conscious that the work we did was not as complete and as satisfactory as it might have been; but we did at least give both types of apparatus a fair trial, and both Warren and I were, if anything, prejudiced in favour of the closed apparatus. Warren has told you what happened when he tried the closed type above the North Col when he was first there. I had not seen him when I got out on the Col with Tilman, Smythe, and Shipton later on, but I heard what happened, and when the opportunity came of going up to Camp 5 I tried the open type first and got surprisingly satisfactory results with it more or less from the word 'Go'. I found no difficulty in acquiring the breathing technique that it demands, and though it did not send me uphill at the speed one might have climbed at in the Alps, it did at least improve my performance as compared with the previous day coming up to the Col. In particular, I noticed a tremendous sense of ease in climbing, in that it took away the great strain that one is apt to feel otherwise. The following day when I made the short test on the closed apparatus I obtained entirely unsatisfactory results, exactly comparable with what Warren had experienced.

The comparison of these two alternative units is a technical matter which I need not bore you with here, but I would submit that the agreed success of the open type of apparatus and its inherent simplicity, which is an enormous advantage under those conditions, and one which those who have never been there cannot appreciate, are very important arguments in favour of it.

If oxygen is to be used successfully on Everest, the design of the apparatus will have to be improved as far as possible—and there is still room for improvement, because the bulk of the weight one carries is not oxygen at all, but metal. Everything possible will have to be done in advance to develop and practise the technique. It is no good taking oxygen as a last resort in case the attempt without it fails.

In appreciation of Tilman's leadership I would like to say that if ever I go to Everest again I very much hope it will be under his leadership.

Mr N. E. ODELL: During the course of an expedition one sometimes falls to wondering what sort of a report will be demanded on one's return home. Lying in my sleeping-bag at Camp 1 towards the end of the expedition I composed a speech befitting, as I thought, such an occasion as this. That particular one was largely forgotten, and I found myself composing another walking in the drenching monsoon rain down the Lachen valley on the way back to Kalimpong. A good deal of the speech so composed was severely critical, for I felt that there was so much that was not as it should be in the running of these Everest expeditions. A swing-over from the large, expensive expedition to the smaller, economical one had by no means been all to the good. True the compacter, smaller party had been a success as a party, and much of our equipment had been reasonably sufficient, but in other respects the cult for lightness and mobility had been carried unnecessarily far. It should be quite evident that the Everest expeditions are, by reason of the position of the mountain and its approach, in quite a different category from those smaller summer undertakings on the southern slopes of other Himalayan peaks. An Everest expedition operates for anything up to six months, from a period of real winter in Tibet to relatively genial summer conditions. Therefore the extreme cutting down of spare clothing and other articles of equipment, as practised on this year's expedition, is not only undesirable but highly risky. I have myself travelled too far in the Arctic and elsewhere not to appreciate what the fortuitous loss of an important garment, for instance, may mean. At the same time in other respects improvisation is often a desirable necessity, but even that has its limits.

But it is mainly in respect of the provisioning that I would criticize this year's expedition. Frankly we did what we did on the mountain not because of our meagre rationing but in spite of it. And we could have done more at times than we did, and returned, some of us, in better condition, if our rationing had been ampler and more suitable. I am no believer in the necessity for truffled quails or champagne (though practically all of us regretted the refusal of a generous gift of a case of the latter), but for a sustained sojourn at really high altitudes a carefully selected and varied diet is essential; and some alcohol has its uses after a particularly exhausting day. As most people know, one's palate and appetite become very fickle at high altitudes, and it is no good thinking that badly cooked porridge, or an inferior brand of pemmican, or a single rasher of bacon for breakfast, is going to keep together the body and soul of even the most devoted Everest climber. Valuable advice in this respect can and should be expected from the medical officer. Moreover, the Shipton-Tilman doctrine of living off the country is definitely inapplicable to, and highly undesirable to try to practise in, Tibet. For one thing, local supplies of food are non-existent or hard to come by, and where purchase is possible from the peasants there is inevitable shortage and likely famine during the succeeding winter. It seems often to be forgotten that one of the great objections of the Lhasa authorities to these expeditions is their tendency to upset the internal economy of the country.

Nor am I at all satisfied that dispensing with a transport officer, as we did this year, has as much advantage as disadvantage. He can become an actual member of the climbing party if necessary, when his duties of transport to the foot of the mountain are completed, as several earlier expeditions have arranged. Certainly the role of interpreter and transport officer should be kept distinct, as was not done this year.

Briefly, in my opinion, a future expedition to Mount Everest will be best advised to adopt a compromise between the lavishness of some earlier expeditions and the frugality of this year's enterprise. I still consider that the conduct of the expedition of 1924 was a model for all time, for if the cost of the unsatisfactory oxygen equipment of that year be deducted the total expenditure was by no means excessive.

I was more than glad of the opportunity of continuing my geological observations which were commenced in 1924 and furthered by Wager in 1933. This work was always carried out as heretofore with strict regard to the susceptibilities of the Tibetans, and most of it was purely observational. As to the stratigraphy of the Everest Series and its actual age, in 1924 I indicated the possibility of its being a southern outlier of Heron's Permo-Trias Series, which outcrops along the southern border of the Tibetan plateau proper. Without further question Wager adopted this suggestion, and from fossil evidence and relationships in northern Sikkim added the very plausible one that the Everest Limestone Series might be considered Permo-Carboniferous or Carboniferous, and not Permo-Trias, in age. If his distant correlation is valid, then by actual local field correlation I was able to determine that the age of the Everest Series is definitely not younger than Permo-Carboniferous. Moreover from those superb view-points of the Lhakpa La and the Rapiu La fresh evidence was obtained as to the actual structure of the Everest and Makalu massif. From Makalu a great thrust-plane runs beneath Lhotse and into Everest itself. In consequence of movement along this thrust-plane the Everest block has been tilted, and the rocks of the mountain given their well-known, and, from the climbers' point of view, unfortunate, northerly dip. Moreover it could be seen that the top of Makalu is composed of a portion of a great sill of light granite whose injection probably accompanied the above movement.

Amongst glaciological observations one's earlier surmises were confirmed, and the Mount Everest glaciers found to be frozen throughout their mass and to have other High Polar characteristics. The Ahlmann glacier-drill was brought into use, but owing to the extreme hardness of the ice little headway could be made with the drilling bits available. The structure of the trough of the East Rongbuk glacier was re-examined, and while my ideas of 1924, as to its being a special zone of compression and shear, were confirmed, it would appear that the blue banding of the ice, then attributed to foliation, is in reality largely due to the intersection of upturned bedding planes of coalescing tributary ice-streams. This intersection, together with the contained moraine and its heat-absorbing qualities, appear to be the main factors in the formation of the East Rongbuk trough, as well probably as in other similar features.

I regretted not being allowed to return via the Kharta valley, for I was particularly anxious to measure some of the remarkable river-terraces of the Chiblung Chu, as suggested by L. R. Wager after his visit in 1933, for that might have tested the idea that the Himalaya may hereabouts have been upwarped since Pleistocene times.

Some meteorological observations were made, particularly determinations of relative humidity with an Assmann psychrometer, kindly lent by the Meteorological Office of the Air Ministry. We had hoped to use a special high-speed anemometer and so obtain for the first time measures of the wind velocity across the face of Everest. But the early arrival of the monsoon, before we had actually reached the North Col, prevented this being done.

I wish here to acknowledge my gratitude for other instruments lent me by the Society for use on the expedition.

The amount of research which can with profit be pursued by members of a Mount Everest expedition should quite eliminate the danger of bedsores, which one of our party this year claims to be the chief menace of such expeditions.

The Deputy President: I feel we should have the opinion of members of previous expeditions. I will first ask Dr Raymond Greene to add a word or two.

Dr Raymond Greene: Most of the remarks I could have made have already been made by previous speakers, but there are one or two points I would like to underline.

First, the question of diet. In spite of what Mr Shipton has said, it is clear that actually there was more minor illness in the 1938 expedition than in the 1933, of which I was a member. I think part of that increase in illness may well be due to the fact that in 1933 we consulted a dietician, and although the food we took was not palatable it was an adequate, well-balanced diet containing the necessary accessory food factors and necessary proportions of the essential food factors. I am quite certain that if Everest expeditions are to remain as fit as possible in the hard conditions in which they live, it is essential that the food should be very carefully considered.

I should like to say as to oxygen that I entirely agree with Mr Tilman that the business of getting up Everest will not be finally settled until somebody has got up without oxygen, and the sooner somebody gets up without oxygen and allows the climbers to go back to climbing for pleasure, the better. Because I have made a special study of oxygen and oxygen apparatus it has fallen to my lot to do a certain amount of work on the subject. Prior to the setting out of the expedition I was asked by the Mount Everest Committee to undertake a certain number of inquiries into the matter of oxygen apparatus. These inquiries were begun months before the expedition started but, as usual, owing to obstructions here and there, the oxygen apparatus, as finally taken, was not properly tested before it went. The reason why the closed circuit apparatus did not work was that it was not quite good enough. If one could design a perfect closed circuit apparatus there is no doubt whatever that climbers could walk up the last part of Everest as easily as they could at sea-level. I have always been in favour of the open type of apparatus, and the one taken in 1938 differed only in minor details from that which I designed for the 1933 Everest expedition. I am not at all surprised, and certainly very pleased, to hear it gave more satisfaction than the closed apparatus.

Lastly, I want to support with all the force at my command what Mr Odell has said as to the scientific aspect of Everest. I think a lot too much has been said in the past about the spiritual and mystical significance of climbing Everest and about its possible effects on British prestige. However true that may be, the fact remains that we go to Everest not for those reasons at all, but either simply because it is fun or in order to satisfy some purely personal and selfish psychological urge.

I do not think we are any longer justified in spending large sums of public money in satisfying these private urges. If it is argued against me that the last expedition did not spend public money and that there is no reason why public money should be spent, my answer is that I do not think even private expeditions are justified in losing the outstanding opportunities which they have for making valuable scientific researches. Not only is it interesting and useful to carry out researches such as those suggested by Mr Odell, but in an age which is becoming increasingly air-minded opportunities should be taken of investigating much more carefully than has hitherto been done the physiology of high altitudes and also the meteorology of the upper atmosphere. I think that future expeditions to Mount Everest should be planned primarily on a scientific basis. I am sure that the necessary funds would be available. The expedition should go out primarily as a scientific expedition accompanied by a small climbing party which, if opportunity arises, will undoubtedly reach the top.

Supposing we go on sending out expeditions which return one after the other but get no nearer the summit than the last, and possibly not quite so far towards the summit, people will get rather tired of Everest expeditions and ask what is the interest of these continued failures. Everybody will know, of course, that they are gallant failures in the sense that no one could have done more than Tilman and his party did to reach the top, but the fact remains that one party after another fails to reach the top. How much more satisfactory it would be if they could come back and say: We did not get to the top, but at any rate we have important scientific achievements to our credit.

I think therefore that we should start straight away to plan a scientific expedition. I do not think we need fear particular opposition from the Tibetans. The scientist the Tibetans object to is a geologist, and there has been a geologist on almost every Mount Everest expedition. A scientific committee to plan the research on Everest should, I am sure, be formed now in order that if and when permission for another expedition comes, that expedition can go out to Everest as a scientific and not as a purely sporting expedition.

Mr H. W. TILMAN: I am in entire agreement with Dr Raymond Greene's suggestion with regard to the sending out of scientific expeditions, except that they should not be sent to Everest. There are plenty of other high mountains which are far more accessible, and in British territory, in which scientific expeditions could do all they want in the way of investigation.

As to Dr Warren's suggestion that only the leader should go on to the glacier and get a sore throat, three out of seven of us this year had colds, sore throats, and coughs long before the glacier was reached, simply through the journey across the country leading to it. While I am open to correction, I think such complaints are far more likely to be the result of over-feeding than under-feeding. At any rate, it is, I believe, generally recognized now in England that coughs, colds, and sore throats are the result of too much to eat.

I must confess I was surprised to hear any criticism of the food, except from Odell, who has not yet finished criticizing the food we ate on Nanda Devi in 1936 and who, in spite of his semi-starved condition, succeeded in getting to the top.

I am sure that Dr Warren as well as Dr Raymond Greene will be interested to hear that we consulted a dietician in 1935, and though Raymond Greene was not there Warren was, and he will bear me out that the food we then had was much more Spartan than that which we had

27*a*. Sherpas on the north ridge of Everest (p. 81)

27*b*. Sherpas on the north ridge of Everest (p. 81)

28. Camp VI (pp. 83 and 89)

on the 1938 Everest expedition. As you know, dieticians have little sympathy with human nature. No two people agree about food, not even husband and wife. But the whole art of travel is to adapt oneself to circumstances. The Everest expedition involves more travelling than actual climbing and, as you have already heard, the difficulty on the mountain is that one does not want to eat any food whatever. In the opinion of at least two of those on the mountain we lived like Sybarites, but I am aware that there are degrees of sybaritism, if one may use the word. Up at Camp 3 we certainly lived very well; above that the choice of food is not particularly vital because of the fact that one does not want to eat at all.

I have already told you what we had to eat, and we found at Rongbuk at least forty boxes of 1936 stores, which some of the party were pleased to see. They contained nutritious things like pickles and liver extracts. I remember as we were about to leave Gangtok the Maharajah gave us a farewell dinner, and very excellent it was. Noticing how wistfully my comrades were regarding the last course of the five-course dinner and having to make a speech in reply to the Maharajah, I suggested that they had all the appearance of men who had undergone a long fast, or were about to undergo one. And I ventured to remind them of a saying by Thoreau, the great apostle of the simple life who lived alone in the American woods and wrote a book about it, that most of the luxuries and nearly all the so-called comforts of life are not only not indispensable but positively a hindrance to the elevation of mankind.

Dr T. G. Longstaff: In moving a vote of thanks to the lecturer this evening I should like, first, to say how much I sympathize with Mr Tilman's views. The large expedition whose objective has been mountaineering has never had any real success in the Himalaya. All the successful Himalayan ascents have been made by comparatively small parties such as Shipton, Tilman, and Smythe are celebrated for. Lord Conway's pioneer expedition was also a small one.

The idea of sending a scientific expedition to Everest is really deplorable; there could be no worse mixture of objectives. It is a district in which all forms of scientific investigation are particularly abhorrent to the Tibetans. If it is desired to send scientific expeditions to deal with terrestrial magnetism and that sort of thing, follow the lead of De Filippi, our lamented member who died a few weeks ago, or Dainelli, or Stoliczka, or any of the eminent scientists who have worked in the Himalaya, and choose a region suitable for the particular work required. Another point to remember is that on Everest expeditions there have always been scientific members. The reason that little really effective work is done by scientists on these expeditions is either due to failure of the individual scientist or to the accuracy of my statement that an Everest expedition cannot properly combine climbing and science.

With regard to oxygen I should like to see it more frankly admitted that little advance has been made on Professor Finch's achievement with

Geoffrey Bruce in 1922. He used a similar apparatus to that employed—on his recommendation—this year. If since that date collaboration with Messrs. Siebe Gorman, the great oxygen experts, had been continuous, instead of invariably spasmodic and at the last moment, a type of apparatus suitable for mountaineers might by now have been evolved. Dr Warren deserves great credit for breaking down custom and trying out apparatus in the Alps last summer.

So, in moving a vote of thanks to Mr Tilman, I say that I think he has proved that a small party has as good a chance as any other of climbing Everest. He did not make it clear that he took two climbing parties of two men to 27,000 ft. in the monsoon, in powder snow, and got everybody away without frost-bite. A very fine performance indeed, really as fine a performance as that of Houston's party on K 2 this year, though that was at a lower altitude.

I beg to offer our thanks to Mr Tilman for the account which he has given of the 1938 Everest expedition.

THE DEPUTY PRESIDENT: We have had an extremely interesting evening, and it only remains for me to ask you to accord the lecturer and those who have discussed his paper a most hearty vote of thanks.

APPENDIX B

Anthropology or Zoology with Particular Reference to the 'Abominable Snowman'

I have here collected and commented upon all the available evidence supporting or denying the existence of this creature. I do not claim to be *the* authority (I know of no one who does), but I happen to have some first-hand knowledge and I have read everything that has been written about it. I shall perhaps be criticized for treating a serious inquiry in a light-hearted way, but it is possible to be serious without being dull. 'Nothing like a little judicious levity', as Michael Finsbury remarked.

Most of the published evidence has appeared not in scientific journals but on the leader page of a respectable daily to which even its enemies will not deny probity and a sense of tradition. Indeed, all who share and respect this last quality must have noticed with regret how the hospitality of those venerable columns were abused by the iconoclasts in a determined attack upon the 'Abominable Snowman', denying its existence or that it ever had existed. No great harm, however, was done. Bear tracks in the course of a column and a half were shown to have been made by bears; wolves, otters, and hares, were found to make tracks after their kind, and when the dust of conflict had settled the 'Abominable Snowman' survived to pursue his evasive, mysterious, terrifying existence, unruffled as the snow he treads, unmoved as the mountains in which he dwells, uncaught, unspecified, but not unhonoured. But I anticipate.

All those interested in the Himalaya and mountain exploration generally are indebted to the first Mount Everest expedition, the reconnaissance of 1921, on many counts; not least because it is to them we owe the introduction of the 'Abominable Snowman' to mountaineering and scientific circles. It is very fitting that a mountain which was for long itself mysterious, whose summit is still untrod, and which has been the scene of many strange happenings, should be the starting-point for an inquiry of this nature. The leader of that expedition, Col. Howard Bury, came across footprints resembling those of a human being on the Lhakpa La, the 21,000 ft. pass north-east of the mountain. In an article which he telegraphed home he referred to this and to the assertion of his porters that they were the tracks of 'the Wild Men of the Snows'. As a safeguard, and in order to dissociate himself from such an extravagant and laughable belief, he put no less than three exclamation marks after the statement; but the telegraph system makes no allowance for subtleties and the finer points of literature, the saving signs were omitted, and the news was accorded very full value at home.

Thus the prodigy was born though not yet properly christened; but a godfather was at hand at Darjeeling in the form of a Mr Henry Newman.

Mr Newman got into conversation with some of these Everest porters on their return to Darjeeling and obtained a full description of the 'Wild Men'—how their feet were *turned backward to enable them to climb easily*,[1] and how their hair was so long and matted that when going downhill it fell over their eyes. The name they applied to them was 'Metch kangmi' —'kangmi' meaning 'snowman' and the word 'metch' Mr Newman translated as 'abominable'. As he wrote long after in a letter to *The Times*: 'The whole story seemed such a joyous creation I sent it to one *or two newspapers*.[2] Later I was told by a Tibetan expert that I had not quite got the force of the word "metch" which did not mean "abominable" quite so much as filthy or disgusting, somebody wearing filthy tattered clothing. The Tibetan word means something like that but is much more emphatic, just as a Tibetan is more dirty than anyone else.' Mr Newman then offered his explanation of the 'Metch kangmi': 'This, I am convinced, is that in Tibet there is no capital punishment, and that men guilty of grave crimes are simply turned out of their villages or monastery. They live in caves like wild animals, and in order to obtain food become expert thieves and robbers. Also in parts of Tibet and the Himalaya many caves are inhabited by ascetics and others striving to obtain magical powers by cutting themselves off from mankind and refusing to wash.'

Here then we have the explanation of how the 'Snowman' acquired his eponymous title and a plausible suggestion as to his reality. For my part I cannot accept such a solution for the problem of tracks I myself have seen or for some of those reported by other travellers which we shall have to discuss presently. We mountaineers may be wrong in thinking that a liking for the high snows is peculiar to us, but I should be astonished to find a native, Tibetan or any other, however guilty, ascetic, or careless about washing, who shared our taste for such places. There are ascetics to be found living not far below the snout of the Rongbuk glacier, but they remain immured in their caves, tended by their admirers, and never in my experience mortify the flesh still further by a promenade up the glacier. In the desire to keep this appendix as short as possible I have selected very rigorously from the large number of letters available, and I notice one from Capt. Henniker, R.E., more for the sake of the charm of the anecdote than for its relevance. The letter from Capt. Henniker (now, I think, Lt.-Col. Henniker, D.S.O.) supported this new theory of Mr Newman's with an actual example, but I cannot accept it as in any way invalidating the refutation of it which I have just given, for a 17,000 ft. pass on a well-known route in Ladakh is not on all fours with the places where strange tracks have been observed, such as the Snow Lake or the

[1] This is interesting because the same backwards technique is sometimes adopted by climbers wearing crampons (ice-claws) when climbing very steep ice or snow slopes.

[2] This may have accounted for the stir created by Col. Howard Bury's laconic statement.

Zemu Gap, both of which are extremely remote from human habitations of any kind or known routes. In 1930 on the summit of a 17,000 ft. pass in Ladakh, Capt. Henniker met a man completely naked except for a loin cloth. It was bitterly cold and snowing gently. When he expressed some natural astonishment he met with the reply given in perfect English: 'Good-morning, Sir, and a Happy Christmas to you' (it was actually July). The hardy traveller was an M.A. of an English University (Cambridge, one suspects) and was on a pilgrimage for the good of his soul. He explained that one soon got used to the cold and that many Hindus did the same thing.

It will be seen after reviewing the evidence which I shall marshal for and against the existence of the 'Snowman' that except for one instance of small value everything turns upon the interpretation of footprints. And if finger prints can hang a man, as they frequently do, surely footprints may be allowed to establish the existence of one. I agree that for some of the many strange tracks which have been reported there is either a definite or a probable explanation, but for others there is not. Not unnaturally the 'Abominable Snowman' always leaves his tell-tale prints in snow which, for obvious reasons, is an unsatisfactory medium. The identification of a photograph of a footprint in snow, taken after the lapse of an unknown number of hours or days after it was made, is no easy matter, and while I do not question the interpretation of the experts I marvel at their confidence. It is true, and it is a pity, that no European has seen or even thinks he has seen an 'Abominable Snowman', but such negative evidence is not really of great value because the number of Europeans who visit his haunts in the course of a year could probably be counted on the fingers of one hand. On the contrary there are many Tibetans and other dwellers near the Himalaya who do claim to have seen him, and we Europeans, I hope, are not to arrogate to ourselves a monopoly of truth. Due weight, too, must be accorded to a tradition which is so very widespread, which covers most of the Himalayan regions from the Karakoram in the west to as far east as the Upper Salween.

Before marshalling the main evidence I must deal briefly with the solitary exception noticed above. In a letter, a Mr H. B. Hudson recounted an experience he had when camping in the Pir Panjal in a glade of pines at about 8,000 ft. 'It was towards evening', he wrote, 'and my servants, all devout Musulmans, were cooking. All of a sudden there was a ghastly yell from among the trees not far away. My servants were obviously terrified as I asked what had made that hideous noise. We sat round the fire discussing what I shall now explain very briefly.' There is no call to follow Mr Hudson's brief explanation for it is merely a recapitulation of native beliefs—Rishis masquerading as bears, spirits of women who have died in childbirth or of persons who have died violent deaths, lonely and longing for human companionship. What a pity Mr Hudson did not take active steps to investigate that 'ghastly yell' instead of posing questions for the Society for Psychical Research. We might have learnt something interesting, although in the terrain he describes—pine trees and a mere 8,000 ft.—I fear it would not have been anything germane to this inquiry. Himalayan travellers are constantly hearing weird noises in far more likely terrain all of which may safely be attributed to the forces of nature acting on ice, snow and rock.

After the instance of 1921 from the region of Mount Everest the first reliable report of strange footprints came from the well-known traveller Mr Ronald Kaulback. Writing about a journey to the Upper Salween in 1936 he reported having seen at a height of 16,000 ft. 'five sets of tracks which looked exactly as though made by a bare-footed man'. Two of his porters thought they were the tracks of snow leopard, two thought they were those of 'mountain men' which they described as like a man, white-skinned, with long hair on head, arms and shoulders. Mr Kaulback added that in those parts there were *no bears*. Corroborative evidence was tendered by Wing-Comdr. Beauman who had seen similar tracks near the source of the Ganges in Garhwal in the Central Himalaya. Various correspondents offered explanations for these tracks, in particular experts from the Natural History Museum. First, they were said to be those of large langur monkeys, large heavily built animals, whose footprints would be little smaller than those of a man. It is not their custom to walk on their hind legs and they are arboreal in their habits, but it was suggested they might traverse the open between one forest and another. To this it was objected that the tracks were seen 3,000 ft. above the tree line, and that although he had spent five months there Mr Kaulback has neither seen nor heard of any monkeys.

The experts withdrew their langur from the competition, the shoe not fitting, and produced their Cinderella: 'We are told there are no bears', they countered, 'but what of the Giant Panda or Snow Bear? Is it not possible the tracks were made by an unknown relative of this species?' The last suggestion was considered a shrewd one by Mr Kaulback who wrote to say he was ashamed he had not thought of it himself. Possibly there was a hint of sarcasm in this admission, since he added a rider to the effect that he had seen no panda-like skins or even heard of such animals in those parts, nor were there any bamboo shoots, a *sine qua non* for pandas without which they languish and die.

So far then we have as candidates for the authorship of queer tracks seen on three several occasions, snow leopards, outlaws, bears, pandas, ascetics, langurs, or *x* the unknown quantity (which we may as well call the 'Abominable Snowman'), roughly in that order of probability. This was the uncertain state of the poll in the summer of 1937, but in the autumn of that year a new and important witness appeared who in one stroke settled or was thought to have settled the matter out of hand. The bear was declared the winner and our 'Abominable' friend was found to be not only at the bottom of the poll but a complete impostor who should never have stood and who must forfeit his deposit.

The evidence which was thought conclusive by himself and others came from Mr F. S. Smythe. In the summer of that year, on a snow pass 16,500 ft. high in the Central Himalaya, Mr Smythe and his Sherpa porters found 'the imprints of a huge foot, apparently of a biped'. Without any ado the Sherpas declared they were those of a 'mirka' or Snowman, but Mr Smythe, with small respect for tradition and less for his Sherpas' zoology, set himself to measuring and photographing the tracks with the calm scientific diligence of a Sherlock Holmes or Dr Thorndike's gifted assistant. So convinced were the three Sherpas, that they volunteered a written statement which I quote in full because it is upon this weak hook that Mr Smythe hangs his innocent victim: 'We Wangdi Nurbu, Nurbu Bhotia, and Pasang Urgen, were accompanying Mr Smythe over a pass when we saw tracks which we *know* to be those of a "mirka" or "wild man". We have often seen bear, snow leopard, and other animal tracks, but we swear that these tracks were none of these. We have never seen a "mirka" because anyone who sees one dies or is killed, but there are pictures of the tracks which are the same as we have seen in Tibetan monasteries.'

To identify the footprints of a thing one has never seen by a picture of what a Tibetan monk imagines it to be is a fairly bold proceeding. Fear sometimes makes men bold and these Sherpas were frightened men who having jumped to an erroneous conclusion felt obliged to stick to it to palliate their fear. Hence their statement and the wild remark about Tibetan pictures to support it. As anyone who has seen a 'thang-ka' will allow Tibetan painting is decidedly post-impressionist. One would as soon expect to find in a monastery an accurate drawing of a battleship as of a 'mirka's' foot.

Mr Smythe's photos were duly developed, as he assured us, 'under conditions that precluded any subsequent faking'; but he need not have bothered, it is not his facts which are suspect but his inferences. The prints were submitted to the Zoological pundits, headed by Dr Julian Huxley, and were by them pronounced (not without some scientific snarling) to be those of a bear, *Ursus arctus pruinosus*. Whereupon Mr Smythe, triumphantly flourishing his Sherpas' affidavit, announced to his expectant audience that 'the tracks described in recent letters to *The Times* were made by this bear, and that a superstition of the Himalaya is now explained, at all events to Europeans'. In short, *delenda est homo niveus disgustans*; moreover, any tracks seen in the snow in the past, the present, or the future, may safely be ascribed to bears. As a *non sequitur* this bears comparison with the classic example: 'No wonder they call this place Stony Stratford, I was never so bitten by fleas in all my life.'

Had it not so happened that the season of 1937 was a very active one in the Himalaya the Snowman's case might well have gone by default.

Stunned by Mr Smythe's authority few paused to reflect that this apparently fatal blow was merely a matter of three frightened Sherpas making a mistake over some obvious bear tracks, and they meekly accepted his assertion for proof. However, the friends of fair play and seemingly lost causes were comforted to see this assertion immediately challenged by one who had just returned from the Himalaya who wrote under the rather unnecessary pseudonym of 'Bhalu'. In the course of an expedition to the Karakoram, while traversing the upper basin of the Biafu glacier (Sir Martin Conway's 'Snow Lake'), he and two Sherpas had seen tracks which, whatever else they might be, were certainly not those of bear. 'They were roughly circular', he wrote, 'about a foot in diameter, 9 in. deep, and 18 in. apart. They lay in a straight line without any right or left stagger, nor was there any sign of overlap as would be the case with a four-footed beast. The Sherpas diagnosed them as those of a Snowman ("yeti" was their term) and they thought he was the smaller man-eating variety and not his larger yak-eating brother. When I pointed out that no one had been in those parts for 30 years and that he must be devilish hungry they were not amused. I was short of film but considering the subject and the suspicious nature of scientists I thought I could spare one. In fact I made two exposures but being less skilful than Mr Smythe made both on the same negative. A few days later in another glacier valley, bear tracks were everywhere and were quickly recognized as such by the Sherpas and myself. They were no more like the others than those of a two-toed ant-eater.'

'Balu's' blundering stupidity with his camera cannot be too much deplored, especially when we consider Mr Smythe's cool, efficient handling of a similar discovery. In default of a photo of the tracks he produced a sketch which roused the naturalists to start a fresh hare, or rather otter. One wrote suggesting that the tracks were those of an otter progressing in a series of leaps. 'The Indian otter (*Lutra lutra nair*)', he wrote, 'has already been reported from high altitudes in the Himalaya.' This hint was snapped up by one brother naturalist who endorsed it and added that he had used it himself some years before to dispose of the Loch Ness Monster; and it was snapped at by another who unkindly pointed out that the otter in question was *monticola* not *nair*, the former being found in the Himalaya, the latter only in Southern India and Ceylon.

Amidst the snarls of the zoologists it was pleasing to hear a modest pipe from one signing himself 'Foreign Sportsman' (strange pseudonym) who introduced yet another small piece of first-hand evidence. He wrote: 'Balu's contribution to the discussion was welcome. His spirited defence of the Abominable Snowman wilting under the combined attack of Mr Smythe and the Zoological Society reminded me of Kipling's lines:

> Horrible, hairy, human, with paws like hands in prayer,
> Making his supplication rose Adam-Zad the Bear.

29. Camp VI, and peak of Everest (pp. 83 and 89)

30. F. S. Smythe at highest point in 1938 (p. 84)

The burden of proof has now been shifted to the shoulders of the Society who must now find us a one-legged, carnivorous bird, weighing several hundredweight. While not wishing to draw a red herring across this fresh line of inquiry may I recount an experience of my own in Garhwal last year. With two Sherpas I was crossing the Bireh Ganga glacier when we came upon tracks made in crisp snow which resembled nothing so much as those of an elephant. I have followed elephant spoor often and could have sworn we were following one then but for the comparative scarcity of those beasts in the Central Himalaya. "Pshaw! A falling boulder!" I hear some grizzled Himalayan veteran exclaim. True, in certain conditions boulders can and do make a remarkably regular series of indentations like tracks, but I have yet to see a boulder of its own volition hop for a mile over an almost flat glacier.'

Wing-Comdr. Beauman then wrote that the subject seemed ripe for investigation on scientific lines. I wrote to support his suggestion, and since the letter recalls the stormy European scene which was the background to this discussion perhaps I may be forgiven for quoting in part: 'Difficult though the world situation is I feel that from the discussion of such irrelevant matters nothing but good can come. For instance, had there been an Abominable Snowman among the exhibits at the Big Game Exhibition in Berlin the visit of Lord Halifax would not have been given the undesirable prominence it has. I notice regretfully that the correspondence appears to be failing and that a zoologist has been afforded space to drive yet another nail into the coffin of our abominable friend having first poisoned him with another dose of Latin. Difficult though it is, the confounding of scientific sceptics is always desirable, and I commend the suggestion that a scientific expedition should be sent out. To further this an Abominable Snowman Committee, on the lines of the Mount Everest Committee, might be formed, drawn from the Alpine Club and the Natural History Museum.'

The correspondence then died, all those who were interested or who had any suggestions to make having had their say. The Aunt Sally put up by Mr Kaulback had been well thrown at but was still there. Many had tried their hand, some of the throws had been pretty wild, and in the end no one had found a foot to fit even one of those five sets of tracks which 'looked exactly as though made by bare-footed men'. The origin of the epithet 'abominable' had been explained, the fauna of the Himalaya well canvassed, and ascetics dragged reluctantly from their caves—but all in vain. Mr Smythe had then put up his own Aunt Sally only to demolish it himself, and when the dust caused by that operation had subsided we found ourselves with two more lots of tracks, in addition to Mr Kaulback's, with no apparent owners.

At that time I had an open mind on the subject with perhaps a slight 'conservative' bias. I should have had no difficulty in concealing my chagrin if the scientific sceptics had been confounded. I had discussed the thing with both 'Bhalu' and 'Foreign Sportsman' and agreed with them that the tracks they had seen were not to be explained away by shouting 'Bear'. The circularity of their tracks was a peculiar feature and in one way significant, for of course a foot so shaped would be ideal for travelling in snow. Bearing in mind that the thing we were looking for was not so much a brute beast as a primitive form of man I propounded to them a theory to fit these round tracks and was laughed at for my pains. Beavers, bees, ants, and some birds, are by no means devoid of constructive ability and we must credit our Snowman with glimmerings of sense. Why should he not have adapted a primitive form of snowshoe? Anyone who has been in country where snowshoes are worn will have seen such a shoe, perfectly round and about 1 ft. across. Such an obvious and simple aid must have been thought of before wheels were known or even before tree trunks were used as rollers for shifting heavy weights. To the objection that since snowshoes are not known to natives of the Himalayan regions I am crediting the Snowman with more than his share of intelligence, my answer is that the people of the Himalaya seldom or never travel on or above the glaciers while the Snowman never travels anywhere else. Necessity, as has been remarked before, is the mother of invention.

To come now to some evidence of my own which to my mind clinches the matter or at the very least clearly shows that for some of these strange tracks an adequate explanation has yet to be found. In maintaining that we really do not know what made some of these tracks the reader will allow that no extravagant demand is being made upon his credulity. I trust neither Mr Kaulback, 'Balu', 'Foreign Sportsman', nor myself will be classed with such masters of mendacity as Baron Munchausen, De Rougemont, or Barrère, for example—the Barrère of whom Macaulay recorded that there may have been as great liars though he had never met with them or read them. It is sometimes easy, of course, to see that which one fervently expects to see or even to persuade others that you have seen it; but these witnesses I have called and others who, like myself, are completely dependent in their travels upon local porters, would as soon establish the existence of the Devil as the 'Abominable Snowman'. For these porters are superstitious to a degree, liable to fits of discouragement, moroseness, or even panic, at any strange sight or untoward happening, and should their employer encourage their superstitions by appearing to share them he would find himself checked and thwarted at every unlucky turn. No, if we cannot wholly believe these men we must at least acquit them of wishing to deceive either themselves or others.

On the way back from Mount Everest in 1938 the party split up. Taking with me two Sherpas I climbed a peak in North Sikkim and then proceeded via Tangu to the Zemu glacier. At Tangu, where there is a rest-house, we fell in with a large party of German scientists led by a Dr Ernst Schaefer who were engaged in a very thorough examination of the fauna and flora and every other aspect of Sikkim. The party included every breed of scientist known to man: ornithologist, entomologist, zoologist, anthropologist, geologist, and other 'ologists of whom I had never even heard. Here, if anywhere, was a team capable of clearing up any difficult problems. I took the anthropologist (an earnest, inquiring man) on one side and over a few glasses of *Kümmel* abjured him to spare no pains in solving the mystery of *Homo odiosus*, and begged him on no account to be put off by the zoologist who would assuredly tell him that any unaccountable tracks he might see in the snow were not those of a 'Snowman', not even a 'Snark', but merely those of a bear.

On 8 July from a camp on the Zemu glacier we set out to make the first crossing of the Zemu Gap, a 19,000 ft. Col between one of Kangchenjunga's southern satellite peaks and Simvu. The weather was thick, the snow soft. Photography was impossible. As we plodded up the long easy snow slope to the Col, crossing the debris of some huge recent avalanches, I noticed by our side a single track of footsteps which, in view of the weather conditions (daily rain and snow) could not have been more than a few days old. The tracks led up the glacier to the Col and then disappeared on some rocks on the Simvu side. I remember feeling rather peeved at the time to think that we had been forestalled by some other climber, and we craned our necks anxiously over the top to see whether the tracks continued down the south side which was extremely steep. They did not, but on returning to Darjeeling, in order to make sure, I began making inquiries. Lunatics are fairly scarce and only a lunatic would go 'swanning' about alone on the Zemu glacier, and had anyone been there the fact would undoubtedly have been known at Darjeeling. To grasp the significance of this evidence it must be understood that it is almost inconceivable that the movements of any climbing party in Sikkim should remain unknown. Porters talk, and leaving aside the number of mountaineering journals and books, the evidence of the letters we have been discussing is enough to show that men who climb in the Himalaya, though they may be strong, are not often silent.

I found that the last visit to the Zemu Gap had been made by a Major John Hunt[1] to whom I wrote and from whom I had the following reply: 'I was in Darjeeling yesterday and had a talk with Renzing (one of the writer's Sherpas) about your crossing of the Zemu Gap. I went up to the Gap in November last year (1937) and you will be interested to hear the following. When we went up there were distinct tracks up the final slope on the Zemu side—I thought at the time that they were both up and down as the tracks were double. From the top, moreover, steps had been cut down the slope on the Talung side to where it ends in the ice cliff. I used them myself to examine the descent. At the time I presumed they were made by the German party (Grob, Schmaderer and Peider) whom we

[1] Now Brigadier Hunt, D.S.O.

had met at Lachen and who had spent six weeks on the glacier before us. I have, however, just received a copy of their book from which it is clear that *they never went to the Gap at all.* What on earth is the explanation of these tracks? They might conceivably have been those of an animal—though most improbable at that time of year with the deep snow we had had—but for the steps cut down the Talung side. Was the Gap crossed earlier last year and by whom?'

To my mind the answer is obvious—by the 'Abominable Snowman'. The writer of this letter implies that this hitherto uncrossed pass was crossed by someone who forgot to mention the fact. No one who swallows that can afford to laugh at those who like myself believe in a more rational explanation. There are degrees of credulity. Major Hunt seems to believe in this anonymous mountaineer who crosses difficult passes single-handed and says nothing about it, but he would probably boggle at the simple and satisfactory explanation of the mysterious tracks which I offer. It is tempting to call the author of them the 'Zemu Gap Snowman'; and such a precise label is justified, for in this case we have two independent witnesses who at different times have seen the same tracks in the same locality. And if one witness is considered biased, the other is a quite impartial observer who has taken no part in the controversy and who for all I know may never have heard of an 'Abominable Snowman'.

This evidence has not been published before so that the scientists have not had the chance to put up candidates from amongst the fauna of Sikkim which by now must be thoroughly well known. (Are there, by the way, any bears there?) But in the Zemu Gap case there were no tracks which could be attributed according to fancy to bears, snow leopards, otters, or gigantic one-legged birds, but plain tracks of large boots. I do not insist upon the boots, because in soft snow it is not possible to distinguish the nail marks which would be proof incontrovertible, but at any rate there were no signs of toe or claw marks; and it is worth noting that my two porters never so much as mentioned the word 'mirka' or 'yeti' but accepted the prints, as I did, for human feet. In fact the first thought that occurred to us, subsequently to be found without foundation, was that one of Schaefer's party had preceded us. And even supposing the tracks were made by boots, as I admit both the Sherpas, myself, and Major Hunt assumed, there is no reason why the maker of the tracks should not have picked up a discarded pair of climbing boots at the old German Base Camp (at the Green Lake on the Zemu glacier) for attempts on Kangchenjunga, where there was, when I saw it, an accumulation of junk of all sorts, the jetsam of several expeditions, and put them to their obvious use. I have hinted that the subject of our inquiry may not be quite so 'dumb' as we think, and we are not to assume that a Snowman has not wit enough to keep his feet dry if they happen to be the shape that fit into boots.

I am unwilling to produce a scale drawing of a Snowman or even number his hairs on the strength of a footprint, though the professors in their search for the 'missing link' are less unassuming. G. K. Chesterton has remarked on the loving care and skill bestowed by them on their building up of Pithecanthropus—a bit of skull there, a few teeth here, and a thigh-bone from somewhere else—until at last they produced a detailed drawing carefully shaded to show that the very hairs of his head were numbered. I merely affirm that tracks for which no adequate explanation is forthcoming (not forgetting the strange Rongbuk stone footprint) have been seen and will continue to be seen in various parts of the Himalaya, and until a worthier claimant is found we may as well attribute them to the 'Abominable Snowman'. And I think he would be a bold and in some ways an impious sceptic who after balancing the evidence does not decide to give him the benefit of the doubt.

APPENDIX C[1]

Use of Oxygen on the Mount Everest Expedition, 1938. By Peter Lloyd

There were two types of oxygen apparatus available to the Mount Everest Expedition of 1938. One was the open type, used by the 1922 and 1924 expeditions, in which the climber breathes a mixture of oxygen and air. The other was a closed circuit apparatus in which pure oxygen is breathed. Now the essence of the oxygen problem is the fact that the apparatus cannot be effectively tested except at very high altitudes. Serious oxygen lack does not make itself felt below about 23,000 ft., so that a test in the Alps is not conclusive even with the closed apparatus, and with the open apparatus it would be useless. Similarly, a test in a pressure chamber is of little value since it cannot include the vitally important acclimatization factor. There was therefore no possibility of making a comparison of the two units in Europe, and it was decided by the leader of the expedition that both must be taken.

The open apparatus was similar in principle to that devised for the 1922 expedition.[2] The oxygen was contained in two Vibrax steel cylinders of 500 l. capacity at 120 atm. pressure, and was fed through an adjustable spring-loaded governing valve to a small canvas bag acting as a reservoir, and thence to the mouthpiece. A pressure gauge was fitted, but no flow-meter, and the governing valve was intended to be set and locked at the required delivery rate. If the lung ventilation of a man doing hard work at 29,000 ft. above sea-level (245 mm. mercury) is taken as 60 l. a minute, then an apparatus of this type delivering oxygen at the rate of 2 l. a minute (as measured at 760 mm. mercury) will raise the partial pressure of oxygen in the inspired air from the original 51–71 mm., corresponding to an atmospheric pressure of 341 mm., or an altitude of 20,800 ft. Similarly, a climber at 25,000 ft. (285 mm. mercury) with the same oxygen delivery and lung ventilation will have an oxygen pressure in his inspired air corresponding to that obtained at 17,500 ft. Since at 20,000 ft. an acclimatized man can still move almost as fast as at Alpine altitudes, it is generally assumed that a delivery rate of 2 l. per minute is about right, and at this rating the apparatus with its charge of 1,000 l. would last a little more than 8 hours. The total weight was 25 lb., the charged oxygen cylinders accounting for 19 lb.

The closed unit was similar to that taken by the 1936 expedition[3] and

[1] Reprinted from *Nature*, vol. 143, p. 961, June 10, 1939.

[2] See Prof. G. I. Finch's account in *The Assault on Mount Everest*, p. 264; also P. J. H. Unna, 'The Oxygen Equipment of the 1922 Everest Expedition', *Alpine J.* vol. XXXIV, p. 235.

[3] See 'The Medical and Physiological Aspects of the Mount Everest Expeditions', by C. B. Warren, *Geog. J.* vol. XC (August 1937).

was of the type which is used for rescue work in mines. It had been improved as a result of tests carried out by Dr Warren in 1937, in the course of which he used the apparatus on the Matterhorn and the Wellenkuppe. The oxygen was contained in a single cylinder of 750 l. capacity, and from the cylinder the gas passed through the reducing valves into a low-pressure reservoir. The user drew oxygen from this reservoir through his face mask, covering nose and mouth, and the expired gases passed back into the reservoir through a canister packed with soda-lime in which the carbon dioxide was absorbed. There were, in effect, three separate automatic valve systems: the pressure-reducing valve which lowers the pressure to about 2 atm. and from which a weep of ½ l. a minute is fed into the reservoir, the breathing valve which opens when the flexible reservoir begins to collapse under external atmospheric pressure, and the lightly spring-loaded mica valves on the flexible connexions to the face-piece, which control the direction of flow through the circuit. All these had to work correctly if the apparatus was to succeed. In addition, there were two hand-operated valves, one the screw-down valve on the main supply and the other a by-pass by means of which the supply of fresh oxygen to the reservoir could be supplemented. As on the open apparatus, a pressure gauge was fitted to indicate the residual pressure in the cylinder.

Every effort was made to lighten the construction, even at some sacrifice of rigidity, but the apparatus weighed 35 lb. for a supply which was calculated to last 5½ hr. This was a weight which a fit man could easily carry over moderately difficult ground, but it might become an impossible burden on steep rocks or if the climber were weakened by lack of food or any other cause. Of the 35 lb. weight, only 14½ lb. was in the charged oxygen cylinder, 10 lb. was in soda-lime and the remaining 10½ lb. was in the frame, the canister and the valve gear. But although the quantity of oxygen was small, virtually all of it could be effectively used, while with the open apparatus most of the oxygen would go to waste in the expired air. It was thought that the high oxygen pressure obtained and the greater efficiency of utilization would outweigh the obvious disadvantages.

A preliminary test with the closed apparatus was done on the journey out to Tibet, at Tangu, in Sikkim (12,800 ft.). In the main, the results were satisfactory, but it was found that in still air the oxygen may get so overheated by the reactions in the soda-lime that conditions become thoroughly unpleasant for the wearer of the apparatus.

The first high-altitude test with the closed apparatus was done by Dr C. B. Warren on the snow slopes above the North Col (23,000 ft.). He found that, although the unit seemed to be in perfect mechanical order, it tended not to stimulate but rather to suffocate him. He was actually moving more slowly than the other climbers, and was forced to stop after every dozen steps to recover his breath.

When the camp on the North Col was reoccupied about ten days later, this gave me a chance for a comparative trial of the two forms of apparatus between Camps IV and V (23,000–25,700 ft.). The open unit was tried first, and it was found that within half an hour the breathing technique had been mastered and the tricks of the apparatus had been learnt. The oxygen was drawn into the lungs through a rubber mouthpiece which was held between the teeth. The technique consisted in biting the rubber to prevent an outflow of gas when breathing out and releasing it when breathing in. When the flow of oxygen was stopped by biting the mouthpiece, the oxygen accumulated in the small canvas bag mounted above the cylinders.

The apparatus was perfectly comfortable. As to speed, I found that I was moving as fast as the fastest member of the party and therefore much better than without the help of the oxygen. But the main difference was the absence of strain or fatigue. The day before, climbing from about 21,000 ft. to 23,000 ft., I had felt really tired, whereas on this day climbing from 23,000 ft. to 25,700 ft. and down again left me comparatively fresh.

Partly to economize oxygen, and partly to test the effect of suddenly removing its stimulus, I turned off the supply at every halt, but this did not seem to produce any reaction whatever. On the other hand, when the first cylinder was running out and I was climbing with a rapidly decreasing supply, the effect was immediately noticeable. The oxygen flow had been set the previous night to a rate corresponding to 2·2 l. a minute at one atmosphere, and at this setting the first cylinder lasted until ten minutes below Camp V. The oxygen supply was turned off 600 ft. above Camp IV and the descent completed without it.

On the following day I took the better of the two closed units and started off with a Sherpa, intending to climb about 1,500 ft., and to compare my times and sensations with those of the previous day. But it soon became evident that there was something very wrong. Mechanically everything was perfect, and the valves were opening and closing like clockwork, but inside the mask I was nearly suffocating, and I had to stop frequently to take a dose of fresh air. If the whole of the system was filled with fresh oxygen, then for a time conditions were pleasant enough, but very quick deterioration followed. The absorption seemed to be satisfactory, for the soda-lime canister was warm—uncomfortably warm in fact—but the effects of the overheating were much more serious this time than they had been under similar conditions at Tangu. The mica valves on the mask were tested and found to be functioning, and the breathing valve controlling the supply to the reservoir was short-circuited by opening the by-pass valve, but nothing that I could do improved the results in any way.

The open apparatus was used again in an attempt on the summit involving three days' climbing, the first day up to Camp V, the second up to Camp VI, and the third for the attempt. It worked perfectly for the second ascent to Camp V, and just over one cylinder was used. On the

31*a*. The Tigers of 1938 (p. 96)

31*b*. Assisting Pasang to Camp IV (p. 95)

32. The Chorten on the track from Rongbuk over the Doya La (p. 66)

following day we set off over heavily snow-covered rocks for Camp VI, which had been pitched at 27,200 ft. Thanks to the oxygen apparatus, which was working perfectly, I was feeling very fit and reached Camp VI half an hour or an hour ahead of the others who were not using oxygen. The camp was magnificently situated on a gently inclined scree slope, and I was interested to find the scree in several places cemented together with ice, which is not known to have been found before at this height.[1] Even here I felt absolutely no ill-effect on turning off the oxygen supply when I stopped climbing.

The story of our attempt to reach the main ridge of the mountain on the following day has already been told. We were forced to retreat after advancing only a short distance above the camp. In spite of oxygen, I seemed to feel the early morning cold as much as my companion Tilman, and I made no more impression than he did on the rocks we were trying to climb. On easy ground, however, it increased climbing speed, as was immediately evident when we were roped together.

It is not an easy matter to do scientific work above 23,000 ft., for Everest has a way of sapping one's energy and leaving little inclination for detailed work. As a result, the tests which were done with the oxygen apparatus are neither so complete nor so precise as one would like them to have been, and our conclusions cannot have the force of proofs. The failure of the closed apparatus in two independent tests on the North Col has yet to be fully explained,[2] but the fact of its failure, coupled with the mechanical troubles which were experienced with it, give powerful backing to the arguments which have already been advanced against it. A closed apparatus was brought back from Everest, and was tested again under actual climbing conditions to ascertain whether there had been any mechanical fault which might have caused the failure. Some repairs were necessary, but the canister was not exchanged, and neither the breathing valve on the flexible reservoir nor the mica valves on the face-piece were touched. The test showed that at these low altitudes (up to 2,000 ft.) the apparatus worked perfectly well.

Perhaps the hardest thing of all is to estimate the advantage conferred by the use of oxygen. Comparison of climbing speeds is difficult and dangerous, but from the trials of 1938, it would seem that from 23,000 ft. to 26,000 ft., the use of oxygen in the open apparatus at 2 l. a minute has only a slight effect on natural climbing speed. The reduction of strain and fatigue, however, provides ample justification for its use at this height. Above 26,000 ft., the increase in climbing speed becomes more and more

[1] In *Everest* 1933, p. 316, L. R. Wager reports the greatest height at which ice was found as 25,700 ft. Some compacted snow (not *névé*, as supposed) was seen at about 28,000 ft.

[2] There is, however, a strong presumption that it was the pressure loss in the breathing valves which, increasing at high altitude, made the apparatus unusable.

apparent. The maximum advantage is obtained, as one might expect, on easy ground where the climber moves with steady rhythm. More difficult climbing requiring greater exertion results in an increase in the rate of breathing, and with the open apparatus this implies a fall in the partial pressure of oxygen entering the lungs.

We suffered in 1938 from the necessity of taking two different forms of apparatus each with different sized cylinders, but it is hoped that the results we obtained, together with the recent work in the field of aviation, will enable a future expedition to make a definite choice.

APPENDIX D

Geological and Some Other Observations in the Mount Everest Region.

By N. E. Odell, ph.d. (cantab), f.r.s.e., f.g.s.

At short notice, and under pressure of much other work prior to proceeding abroad, it is only possible to give a brief summary of geological and some other scientific observations which were made during the Expedition of 1938, as well as a few comments on important work that is yet to be done. In the course of operations, primarily concerned with the ascent of the mountain, some opportunity was given to continue the geological work commenced by me in 1924 (7, 8), and continued by L. R. Wager during the Expedition of 1933 (14). At the close of the abortive attempt to climb the mountain, it must be emphasized that much further valuable observational work could have been carried out in 1938 had reasonable facilities and a sympathetic attitude been forthcoming on the part of the leadership of the Expedition.

It is unnecessary here to repeat what has already been said in the scientific records of the earlier expeditions to Everest, and for details the reader is referred to the numbered list of literature at the end of this Appendix. Suffice it to say that Sir Henry Hayden (on the Lhassa Expedition of 1903–4 (1, 3), and A. M. Heron (of the Geological Survey of India) on the Reconnaissance Expedition of 1921 (1, 4, 5), first mapped geologically the Tibetan districts lying to the north of the main Himalayan chain and the Everest group; but the stratigraphy and structure of the central mountainous area were left to their successors.

In 1924 a detailed study of the crystalline complex of the main range in particular was commenced by myself, and I published a provisional geological map of the district surrounding Everest itself (7). Wager amplified this map, and drew some important geological sections through the region (14). Structural and petrological work was continued by myself in 1938, and an extensive collection was made of the various rock types which are representative of the metamorphic series of the district. It is a tragedy that on the return journey practically the entire collection was stolen from our camp at Tinki Dzong by Tibetan natives, the theft being presumably in mistake for the cash-box. But an even worse tragedy has been the loss in 1941, on a convoyed voyage to India of all the writer's field-notes, diaries, maps and partially completed papers, relative to work in the Himalaya covering fourteen years. Such a shattering blow, even if not responsible for actual mental unbalance, nevertheless, can scarcely be considered conducive to a sweet and mellow attitude of resignation to events of the late war! In any case, it will be a most difficult and palpably deficient task, if and when the writer is eventually able to make good from memory, and with the assistance of Professor Wager, some of this disastrous loss.

Briefly, it may be reiterated that the southern portion of Tibet, lying north of the eastern Himalaya, consists principally of Mesozoic rocks, which are represented by Jurassic shales and quartzites that have suffered on the whole only a moderate degree of folding. But the southern border of these rocks rests upon a series of limestones which, in spite of certain sporadic fossil-finds made by Heron, have until lately been of doubtful, but possibly Permo-Trias, age. These massive limestones in turn overlap on to the mixed crystalline rocks of the main Himalayan axis.

The crystalline and metamorphic rocks of the Himalayan zone, including the Everest massif, consist in the main of altered slates, quartzites, siltstones, and thick (often impure) limestones. Everest itself and its satellites are carved out of a thick series of slaty rocks, schistose siltstones and crystalline limestones, in which occur fortuitously veins of fine granite (aplite and pegmatite), containing sometimes pink garnets and small black tourmalines. The topmost block of Everest is composed of dark impure limestone, now recrystallized to form marble. The point to be emphasized here is that all this mass of rocks, which form the higher portions of the main chain, is of sedimentary material, once laid down at sea-level, and is not, as is so often suspected by the layman, made up of granite or other igneous material, nor yet of volcanic rock (13). Granite does occur at places deep down in the range, and veins are sometimes to be found extending from it into portions of the overlying sedimentary series. Another point of interest, which cannot here be expanded, is that these latter rocks are derived from the sediments once laid down along the shores of the ancient Tethys Sea which, in periods prior to the rise of the Himalaya, extended eastwards from the Mediterranean into what is now Tibet.

The problem of the dating, even approximately, of these sedimentary rocks which form the bulk of Everest and the neighbouring mountains has been no easy one owing to all fossils having been destroyed by the pressure, and partly the heat, to which they have been subjected. In 1924 at an altitude of about 25,500 ft. on the north ridge of Everest I found what appeared to be fossil forms, which were localized in a bed of metamorphosed calcareous sandstone some 4 ft. in thickness. However, microscopic examination of these supposed fossils led to the conclusion, which was supported by Mr R. B. Newton of the British Museum (Natural History), that they were not due to organisms but represented an unusual and curious instance of 'cone-in-cone' structure. The latter can be considered the consequence of compression and shearing in the rocks; although an alternative theory, held by some geologists, points to this structure, where it is developed in other cases, being due to settling and volume-shrinkage during the slow de-watering of highly saturated and loosely packed sedimentary materials.

In the absence of fossils, another clue presented itself to the writer in 1924, when it occurred to him that the sedimentary mass of Everest might represent, at any rate in its upper limestone elements, an 'outlier' of the series of limestones (Permo-Trias?) which extend in places along the southern border of the Jurassic rocks of Tibet, as cited above. Wager, from his observations in 1933, adopted this view. But by Wager's own meritorious field-work (16, 17), as well as the palaeontological collaboration of Drs H. M. Muir-Wood and K. P. Oakley (6) of the British Museum (Natural History), it has now been reasonably established, through correlation of certain fossiliferous rocks (Lachi Series) in northern Sikkim with the above limestones, that the latter are carboniferous to Lower Permian in age. Moreover, from his study of the Lachi fossils Oakley states (p. 70): 'it is reasonable to suppose that the Mount Everest Limestone falls within the Carboniferous system.'

While the above evidence is of great interest and importance from the standpoint of the actual age of deposition of the rocks themselves, it is of course quite irrespective of the date of uplift of this part of the Himalaya, and of its mode of elevation. As in the case of the Alps, so with the Himalaya, the movements have been a long drawn-out process, during the Tertiary period. There appear to have been three important phases of the upheaval of the whole mountain system. The first occurred towards the end of the Eocene epoch, culminating in the Oligocene, say some 50 million years ago. During this phase the central axis of ancient sedimentary and crystalline rocks was ridged up. The next phase of greater intensity took place about the middle of the Miocene period. The central part of the range, as well as the outer foot-hill districts of the Siwalik deposits, were elevated during the last phase which was chiefly of post-Pliocene age, but did not cease until after the middle of the Pleistocene. There are, however, certain indications that elevatory movement may have been slowly continuing until the present time. These matters are discussed in greater detail by Sir Henry Hayden (1), and by Mr D. N. Wadia in his book *Geology of India*.

As to the nature of the movements which went to form the great Himalayan chain there has been a great deal of discussion amongst geologists, and these the writer has reviewed in a recent paper (13). Although locally in the Simla district and in Garhwal it has been shown by W. D. West and by J. B. Auden respectively, of the Geological Survey of India, that overthrusting on an Alpine scale seems to have taken place, yet there is in my opinion insufficient evidence as yet forthcoming from the eastern part of the range to warrant the hypothesis of far-travelled rock sheets (*nappes de recouvrement*). Wager, however, following Dyrenfurth (2), inclines to such a view in explanation of the structure of the Sikkim area and the tectonic position of the Darjeeling gneiss (Fig. 3 in (14)). In the Everest district Wager and I are agreed as to the evidence predominantly of compression and shear in having reduced the local shales,

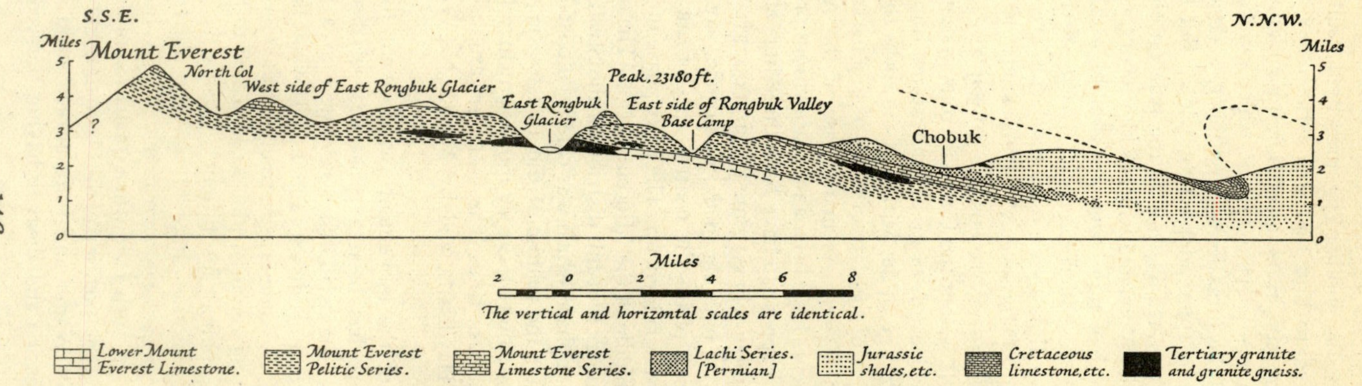

Fig. 1. Section through Mount Everest and Chobuk (L. R. Wager)

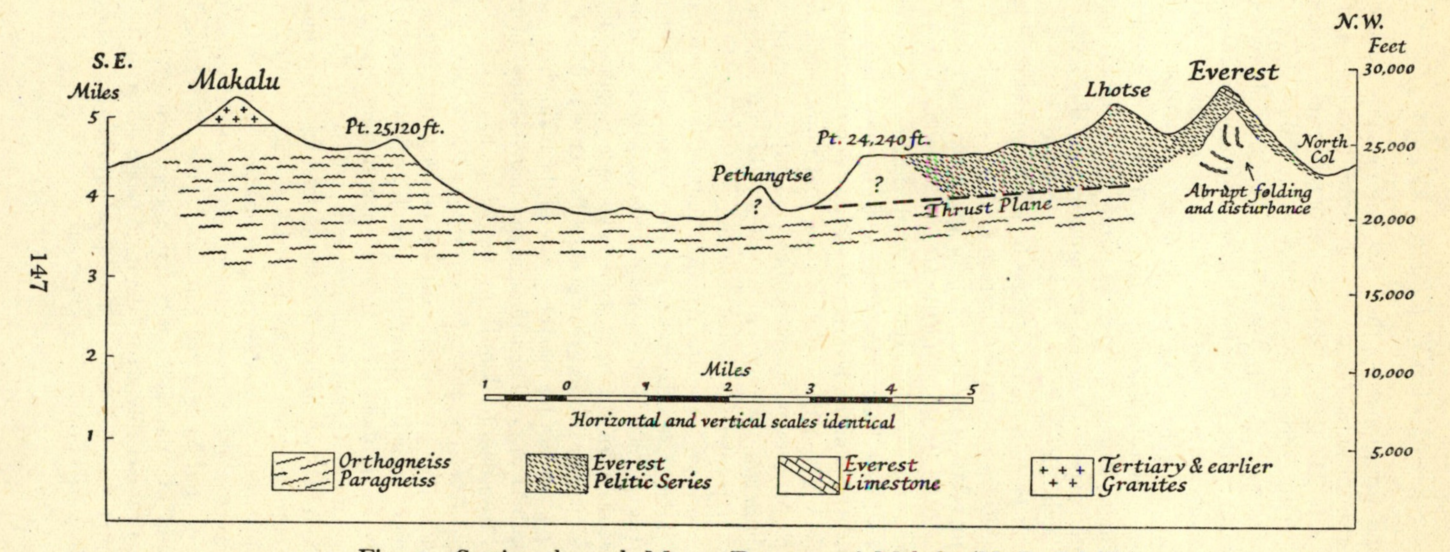

Fig. 2. Section through Mount Everest and Makalu (N. E. Odell)

quartzites and limestones to their present metamorphic condition. In places, too, granitic veins (aplite or pegmatite) have no doubt contributed to this condition. The general disposition of the series is admirably shown in Wager's section through Mount Everest and the region to the north, here reproduced as Fig. 1 (p. 146). There appears to have been a general lateral squeezing, relief from which has been obtained by intimate shearing within the mass rather than by folding or extensive thrust-faulting. 'Lit-par-lit' injection has in places accompanied this shearing. But for the theft of my petrological specimens collected in 1938 (as cited above, p. 143), a good deal more information on the nature of the forces operative upon, and the metamorphic changes wrought within, this interesting series might have been obtained.

However, in 1938, from those superb view-points of the Lhakpa La and the Rapiu La, I was able to obtain fresh evidence as to the actual structure of Everest in relation to the Makalu group to the southward. There seemed to be little doubt from the structural relations here visible that a well-marked thrust-plane runs beneath Lhotse (the south peak of Everest) and into the base of Everest itself. In consequence of movement along this thrust-plane the Everest block has been tilted, and the rocks of the North Face given their well-known, and notoriously inconvenient, northerly dip. Moreover, by marked colour-contrast, it appeared that the top of Makalu is composed of a great sill of light granite, whose injection may have accompanied the above thrusting movement (Fig. 2, p. 147).

Whilst there have been lateral movements of the earth's crust, acting either from Tibet towards India, or vice versa, to build the piled-up columns of strata which form the main ranges, there is evidence that later vertical movements of the block type have been responsible for much of the present elevation of the great peaks. Immense fault-scarps are to be found along portions of the northern side of the main chain, cases in point being Chomolhari in northern Bhutan and many of the mountains along the northern frontier of Sikkim. As the writer has stated elsewhere (13): 'it is to these local elevatory forces of recent date, acting in excess of general denudational processes that the supreme height of the Everest and Kangchenjunga massifs, standing 6,000 ft. approximately above the average of their neighbours, can alone be due: nothing else can explain the often fresh and unweathered mountain walls, which rise sheer to heights of 6,000–10,000 ft. above their bases, and extend for many miles round the massifs.' Whether or not we are to attribute such vertical movement to isostatic adjustment, consequent upon loading and unloading of the Himalayan tract in the course of its development and erosion, cannot be discussed here; but Wager has put forth some interesting ideas in this connexion and in relation to the drainage pattern of the Arun River, which cuts so remarkably across the main range between the Everest and Kangchenjunga massifs, and for this question reference should be made to the original paper (15).

Glaciology

Under this heading there is not a lot to add to what has already been described by me in 1925 (7), and Wager in 1934 (14). Under the prevailing climatic conditions, differing widely from those of the temperate zone, the glaciers of the Everest region appear to be mainly frozen throughout their mass and have other characteristics of the 'polar' type. Very little melting beyond what takes place superficially at the surface is in evidence, and then only during the warmer months. This perennial frozen condition would seem to give rise to a greater brittleness of the ice, and a tendency, as in the polar regions, for the glaciers to move more by fracture and shear than by what approximates to viscous flow, which in the case of temperate glaciers is due to their being at melting-point. It had been hoped in 1938, by means of an Ahlmann glacier-drill, to obtain further data on the range of temperature changes within the ice, but owing to the extreme hardness of the ice and the unsuitability of the drill-bit little progress could be made. The glacier-drill, however, if it served no better purpose, at least provided much opportunity for caustic jesting from our notoriously anti-scientific leader!

Observations upon the remarkable 'trough' of the East Rongbuk glacier were carried a stage further. While my ideas of 1924, as to its being a special zone of compression and shear, were confirmed, it would appear that the blue banding of the ice, then attributed to foliation (as in rock masses) is in reality mainly due to the intersection of upturned bedding planes of coalescing confluent ice-streams. This intersection, together with the contained line of medial moraine and its heat-absorbing qualities, would seem to be the main factors in the formation of the East Rongbuk trough, and some others in neighbouring glaciers. Foliation, however, as displayed in blue banding, occurs at times quite independently of bedding structure (see Pl. 33 *a*).

As to the giant pinnacles of these glaciers, which are usually to be found associated with troughs, many of them are formed, as stated elsewhere (7, 8), by ablation-sculpture of portions of the ice-walls along the sides of the troughs. Moreover, they are in no way to be regarded as *séracs* due to crevassing. As the writer has described previously, there is much to suggest with Wager (14) that these pinnacles are initiated as hummocks on the irregular *névé*-fields, free of moraine, above the level of commencement of the troughs. Under the prevailing conditions of high altitude, wastage (ablation) of the snow and *névé*-fields takes place, not by melting but by evaporation alone. On account of the low latitude and high solar radiation the development of hummocks and hollows becomes greatly emphasized. From these there eventually grow 'sun-spikes' and 'sun-cups' and pits, of all sizes, according to the local conditions, a structure of high-altitude snow- and ice-surfaces, which was first described as 'nieve penitente' from occurrences in the Andes of Argentina. My further observations of 1938 went to show that in many instances, if not all, out of growing hummocks and other eminences there were progressively developed the huge pinnacles, which in the course of their lengthy journey with the moving glacier, sometimes attained a height of over 200 ft. above their bases. Moreover, it was observed that the surfaces of these giant pinnacles were frequently covered with secondary sun-cups and pits. Most of the pinnacles, too, are inclined ('penitent') with steeper side to the southward, a feature consequent upon their latitudinal position, viz. the particular local declination of the sun.

That such pinnacle-structure is not confined to glacier-surfaces only is shown by similar effects on snow-fields elsewhere. On the Pang La, 18,500 ft., and on alpine meadow-lands at an altitude of about 16,500 ft. in the Kharta valley, were seen extensive areas on which spikes and blades, up to 2 ft. 6 in. high with associated sun-cups or sun-pits, were oriented in an east and west direction (see Pl. 33 *b*). The spikes and blades had a slight inclination towards the southward, like the larger penitent forms of the glacier. Moreover, it was noticeable that no melt-water at all was present in cups or pits, nor yet any trickling away from the snow surfaces. The prevailing diurnal air temperatures being well below freezing-point, until after their general dissipation, all ablation is due to evaporation, and not ordinary melting. This point is of interest, since it implies that extensive snow-fields and glacier-surfaces can suffer wastage at high altitudes without the formation of melt-water and all its consequences for the denudation of mountain flanks and lower valleys. It is an important economic consideration, too, in regions which may be mainly dependent on melt-water from snow-fields and glaciers for human supply, whether for consumption, irrigation or power purposes.

A further matter of interest and importance is that of the probable general recession of the glaciers in common, perhaps, with those of the Temperate Zone throughout the world. The glaciers of the Everest massif appear at present to be in a state of equilibrium, supply of snow balancing wastage of ice. It is possible, however, and indeed probable, that they are in extremely slow retreat. That they once extended much farther down their valleys is clear from the evidence of moraine-shelves along the Rongbuk valley and elsewhere. These consist of four lateral moraines, or discontinuous portions of them, one above the other, occurring at heights of from 400 to 800 ft. above the valley-floor. The shelves, together with the marked recessional moraines to be found at intervals along the valley-floors, indicate halting stages in the general retreat. There is insufficient evidence forthcoming at present to suggest separate periods of glacial extension as has been demonstrated elsewhere in the Himalaya. Moreover, it is uncertain whether the glaciers during the maximum stages of the Pleistocene debouched far out on to the Tibetan plains or not. On the northern side of the main range we know, and Wager has emphasized, that they extended a mere five or six miles

33*a*. Pinnacles of Rongbuk glacier, 1938—showing seasonal bedding (curved) and foliation (vertical)

33*b*. Sun-blade structures, Kharta Valley, Tibet, at 16,500 ft.

beyond their present snouts (14), but there is insufficient evidence to show that they ever joined up to form a Tibetan ice-sheet, a view that was held by Blandford and supported at one time by the writer (8). It must be remembered, however, that there is widespread evidence on the southern flanks of the Himalayan chain of glaciers having extended at the culmination of the Glacial epoch, perhaps 80 miles in some cases, farther down their valleys than the present position of their termini.

As to the study of the widespread Post-Glacial gravels and silts which stretch far down the valleys and along the borders of the Tibetan plains, as well as the extensive terracing to be found in the drainage basin of the Arun River, little beyond passing observation has yet been accomplished. An immense field, therefore, lies open for further research. Wager has discussed the interesting evidence that some of these valley-terraces are out of level, or warped (15), and has pointed out the significance of this fact in relation to the further important one of the continuing uplift of the Himalaya, which is so often postulated. It was a matter of the greatest regret that I was prevented after the climbing operations in 1938 from continuing Wager's work upon this aspect of the tectonic history of the region. In the valley of the Chiblung Chu and elsewhere, however, data were collected which suggest that much of the terracing, as well as torrential deposits in those areas, may well have been due to volume increase of melt-water (climatic rejuvenation) consequent upon Late- and Post-Pleistocene conditions rather than to tectonic rejuvenation or other orogenic change. On the other hand, when at the end of April we retreated from Everest to our rest-camp near the Arun gorge, I noticed at the confluence of the Kharta Chu and the Phung Chu that the incision of its meanders by the latter river to a depth of 50 ft. suggested uplift of the area as the more likely cause hereabouts. Since not only the main stream of the Phung Chu, which becomes the Arun River, but all three of its local tributaries from Kharta, Lang and the Doya La have notably entrenched themselves, it would seem reasonable to suppose in this case rejuvenation by uplift of the region to the northward, or alternatively relative depression to the southward.

Geophysics

A word should be said in regard to this important modern handmaid of the geological sciences, whose aid in many problems of the uplift of the Himalaya, as well as the physical condition of Tibet, geologists would assuredly crave. A few isolated gravity determinations at the north-western end of the chain have shown that in common with other great mountain ranges of the world there is a deficiency of gravity (i.e. negative anomalies) in that part of the Himalaya. What is sorely required is a systematic network of gravity determinations all along both sides and within the main chain. I had hoped that a start might be made at the eastern end of the range during our journey to Everest in 1938, but my attempts to arrange for a small *bandobast* in this respect met with no response, only cold discouragement. As far as the high plateau of Tibet is concerned, that outstanding problem of physical geology and geophysics, we are as yet in complete ignorance regarding its state of gravity and the distribution of the probable anomalies. Even a few determinations of the existing force of gravity on the plateau would help us the better in understanding how that immense tract remains in such an elevated position without, apparently, the support inherent in folded mountain ranges. For the Himalaya themselves and the Trans-Himalayan ranges on the one hand, with the Kun-Lun chain to the northward on the other, cannot be supposed to constitute, as some might suggest, adequate buttresses for the shoring up of thousands of square miles of plateau. A couple of trained observers, with a modern light gravitational instrument, could accompany another Everest expedition and do invaluable work. Moreover, they could at the same time carry out observations on terrestrial magnetism which should be of great interest and value, and would amplify the important results of this kind which were obtained in Sikkim and Tibet in 1938–9 by Dr Karl Wienert of the (German) Schaefer Scientific Expedition. These geomagnetic data are shortly to be published in full, and are, I gather, about the only results of this Expedition to survive the war. In a recent letter Dr Wienert states that his terrestrial magnetic survey, amongst other things, indicates that the Tsangpo valley is a tectonic line, like the valley of the Upper Rhine between Schaffhausen and Mainz. For all its Nazi flavour and its S.S. guile, this expedition at least did good work in this respect, and moreover was able to link up with the great magnetic traverse across Central Asia by W. Filchner, which took twelve years to carry out.

Meteorology

Apart from the usual temperature and other meteorological observations, particular attention was paid in 1938 to determinations of relative humidity with an Assmann psychrometer, kindly lent me by the Meteorological Office of the Air Ministry. In the light of the findings of Mr Gerald Seligman that under Alpine conditions 85 per cent. relative humidity constitutes a critical point at which wind-slab avalanches may take place, an attempt was made to test the theory under high Himalayan conditions, and particularly on the slopes of the North Col. Under the difficult conditions of 1938, however, no satisfactory conclusion was reached. It should be emphasized that this is a most important piece of work for a future expedition, even if an ordinary dry and wet-bulb thermometer only, much less precise than the Assmann psychrometer, is available.

Moreover, it had been hoped to bring into use a special high-speed anemometer and so obtain for the first time quantitative values of the wind velocity across the north face of Everest and elsewhere. But the exceptionally early arrival of the monsoon, before we had actually established ourselves on the North Col, unfortunately prevented this being done.

It is a further tragedy, and loss to science, that such meteorological results as we obtained in 1938 were lost in the course of the late war.

In regard to the important practical matter of the arrival of the first monsoon snow on the face of Everest, it was suggested to me in India by Dr (now Sir Charles) Normand, then Director of the Indian Department of Meteorology, that much information might be gleaned from season to season by mountaineers making distant observations from the peaks of northern Sikkim, if climbing in that country just prior to the monsoon. Such observations might prove of value in arriving at some conclusion as to the frequency of such unusually early precipitation of monsoon snow as in 1936 and 1938, and the remarkable phenomenon of such precipitation occurring upon the Everest group well in advance of the establishment of monsoon conditions in peninsular India. It would appear that upper currents of an unusual character and intensity may be responsible for thus skipping over India in the earliest phases of the monsoon.

Archaeology

Along the route through Southern Tibet one sees relics of many ruined villages and erstwhile mud-walled towns, different in style of architecture and construction from present-day buildings and without doubt in many cases dating back hundreds, if not thousands, of years. We found on one site some pottery of quite an unusual type, whose provenance is indeterminate. Many of the buildings suggest fortifications, and there appears in some instances to be much in their character to recall the 'limes' (forts) of eastern Sinkiang, as described by Sir Aurel Stein. The study of these would clearly yield results of great ethnographical and historical value, and amplify the work of Sven Hedin and others in the area north of the Tsangpo River.

It is unnecessary to mention here the large amount of important ecological and other work in natural history which remains to be done throughout the region traversed by an expedition to Everest. Wollaston, Longstaff, Hingston and Greene made considerable collections, but some of their material was lost. Great opportunity is still afforded for valuable work in observation and collecting by specialists or by climbers, who have no reason whatsoever to complain of boredom, bed-sores, or 'unemployment' in the course of these expeditions! A climber becomes, indeed, rather a pathetic if not unreasonable figure, who is given the privilege of travelling through an unusual and beautiful country like Tibet, and who, although a so-called educated man, has no eyes for the outstanding and fascinating wonders of nature on every hand, whether at lower or higher altitudes, be they living or lifeless.

The objection of native prejudice, and resentment particularly of collecting work, that is in some quarters brought forward against scientific work on these expeditions, has often been unduly exaggerated, especially in regard to geology. For actually, at various places in Tibet, both lay and ecclesiastical members of the community have sometimes offered fossils and other odd rock specimens for sale to members of the expeditions. With due tact and appropriate care it is quite easy in the pursuit of one's investigations to avoid offending the more unenlightened elements of a people who still largely live in an atmosphere of the Middle, if not Dark, Ages, and who have cogently declared that Western civilization can do nothing for them but promote unhappiness.

SELECT LIST OF LITERATURE ON THE GEOLOGY AND GLACIOLOGY OF EVEREST

1. S. Burrard, H. H. Hayden and A. M. Heron (1933). *A Sketch of the Geography and Geology of the Himalaya Mountains and Tibet*, 2nd ed.
2. G. O. Dyrenfurth (1931). *Himalaya—Unsere Expedition*, 1930. Berlin.
3. H. H. Hayden (1907). The Geology of the Provinces of Tsang and Ü in Central Tibet. *Mem. Geol. Surv. India*, vol. xxxvi, pt. 2.
4. A. M. Heron (1922). Geological Results of the Mount Everest Reconnaissance Expedition. *Rec. Geol. Surv. India*, vol. liv; and *Geogr. Journ.* vol. lx.
5. A. M. Heron (1922). The Rocks of Mount Everest. *Geogr. Journ.* vol. lx.
6. Helen M. Muir-Wood and K. P. Oakley (1941). Upper Palaeozoic Faunas of North Sikkim. *Palaeontologica India*, vol. xxxi, Mem. 1.
7. N. E. Odell (1925). The Rocks and Glaciers of Mount Everest. *Geogr. Journ.* vol. lxvi.
8. N. E. Odell (1925). Geology and Glaciology. Appendix to *The Fight for Everest*, by E. F. Norton. London.
9. N. E. Odell (1926). Supposed Fossils from North Face of Mount Everest. *Quar. Journ. Geol. Soc.* vol. lxxxii.
10. N. E. Odell (1938). Mount Everest Expedition, 1938 (Tilman, Discussion). *Geogr. Journ.* vol. xcii.
11. N. E. Odell (1939). Some Scientific Researches in the Region of Mount Everest. *Nature*, vol. cxliii.
12. N. E. Odell (1941). Ablation at High Altitudes and Under High Solar Incidence. *Amer. Journ. Sci.* vol. ccxxxix.
13. N. E. Odell (1943). The So-called 'Axial Granite Core' of the Himalaya: its Actual Exposure in Relation to its Sedimentary Cover. *Geol. Mag.* vol. lxxx.
14. L. R. Wager (1934). A Review of the Geology and Some New Observations, Ch. vii in *Everest* 1933. London: Routledge.
15. L. R. Wager (1937). The Arun River Drainage Pattern and the Rise of the Himalaya. *Geogr. Journ.* vol. lxxxix.
16. L. R. Wager (1939). The Lachi Series of North Sikkim and the Age of the Rocks forming Mount Everest. *Rec. Geol. Surv. India*, vol. lxxiv.
17. L. R. Wager (1942). Permian Fossils from the Eastern Himalaya. *Nature*, vol. cxlix.

INDEX